Mandela Effect

A Beginners Guide to the Rising Phenomenon

(The Epic Conclusion to the Worldwide Phenomenon)

Alice Mizelle

Published By **Ryan Princeton**

Alice Mizelle

Mandela Effect: A Beginners Guide to the Rising Phenomenon (The Epic Conclusion to the Worldwide Phenomenon)

ISBN 978-1-77485-721-2

No part of this guidebook shall be reproduced in any form without permission in writing from the publisher except in the case of brief quotations embodied in critical articles or reviews.

Legal & Disclaimer

The information contained in this ebook is not designed to replace or take the place of any form of medicine or professional medical advice. The information in this ebook has been provided for educational & entertainment purposes only.

The information contained in this book has been compiled from sources deemed reliable, and it is accurate to the best of the Author's knowledge; however, the Author cannot guarantee its accuracy and validity and cannot be held liable for any errors or omissions. Changes are periodically made to this book. You must consult your doctor or get professional medical advice before using any of the suggested remedies, techniques, or information in this book.

Upon using the information contained in this book, you agree to hold harmless the

Author from and against any damages, costs, and expenses, including any legal fees potentially resulting from the application of any of the information provided by this guide. This disclaimer applies to any damages or injury caused by the use and application, whether directly or indirectly, of any advice or information presented, whether for breach of contract, tort, negligence, personal injury, criminal intent, or under any other cause of action.

You agree to accept all risks of using the information presented inside this book. You need to consult a professional medical practitioner in order to ensure you are both able and healthy enough to participate in this program.

Table Of Contents

Chapter 1: Mandela Effect

The MANDELA EFFECT is a result of the occurrence of more and more instances in which memories and the reality of things are challenged. This is because there is a disconnection between what people believe from a person's memory and what is proven to be the truth. Mandela effects can also be interchangeably described as quantum changes as well as Mandela changes. Every one of these disconnections from reality generally impacts a large number of people. However, each individual is affected by an inconsistency collection of Mandela effects.

A sign that an aspect is experiencing an Mandela change is the presence of a version in the present as well as an alternate version of it that is cherished by a particular group of people. There is a broad range of topics affected by this bizarre phenomenon, including movies, songs and logos for companies. These changes are usually minor aspects that have no impact on people who do not notice the change. The people who have observed a change experience reactions that range from denial an insistence that something has changed.

The person who is the source of Mandela's effect is Mandela influence is former South African president, Nelson Mandela. This phrase was created by Fiona Broome, who (falsely) believed Nelson Mandela's death took place in the 1980s , while being in prison. It was noticed enough to give a name to this phenomenon when she found out that others attending a conference had also been thinking of Nelson Mandela's passing in the same manner. In reality, Nelson Mandela survived his time in prison, was elected president in 1994 and passed away in 2013. The people affected by the Mandela effect can even recall the news coverage about the death of Nelson Mandela, including the speech he delivered by the wife of his. The thing that is unique about this particular instance is that the group is more than 20 years out of date when it comes to comparing their memories with the reality. This case is a ludicrous example due to the fact that the gap between the death of Nelson Mandela during the 80s, and in 2013 could have had a significant influence on the current historical development in South America.

The most popular explanation to explain the memories linked with the Mandela effect is based on the limits of our brains. This is an excellent argument against any theory that relies on the person's memory, given that this explanation seems plausible, and many people can understand the apprehension of this shortcoming. Many of the memories linked to Mandela changes stem from things that people haven't paid any focus on since their childhood. This puts the flaws of memory in the spotlight because memories from childhood are especially susceptible to changes. People who believe that Mandela effects are due to flawed memories that are frequently called "skeptics" as well as "debunkers" of alternative explanations.

Alternative explanations are generally in comparison to the logic of people who have misunderstood a simple aspect. The most straightforward explanation is usually the most preferred solution to a question, so it's not a surprise that many people gravitate toward this explanation. There is also a wealth of studies that prove the extent to which memory is.

One of the main arguments that supports quantum effects arising from the brain's limitations is the prevalence and existence in the presence of fake memories. False memories are untrue recollections that are caused by a variety of factors, such as implanted concepts and the subconscious overwriting of memories that are in conflict with the concepts shared by members of the peer group. The presence of false memories is beneficial as people are social beings who's brains are active in conforming to external influences. The stability and unity provided from false memory is superior to the conflicting thoughts that coexist within the mind of a person. The level of trust associated with false memories makes them difficult to identify and accept as being incorrect.

This is unfortunate because people are less likely to conduct studies and assess the legitimacy of opposing opinions when their memory is backed by a high degree of confidence. The challenge of confronting someone with false memories they're already familiar with, like those that are associated with Mandela effects is a difficult task.

False memories can be facilitated due to the fact that our brains store information using a proximity-based organizing method that is based on similarity. As memory diminishes because of wear and tear, it's possible for like memories to merge. The new ideas may replace the weaker memories and could appear as more clear memories due to being made up more recent. People who have poor memories for the exact concept might experience a similar mixing of ideas via an external source like leading questions, which are frequently employed during Mandela quizzes. This could result in the possibility of a group of individuals experiencing an unshared memory that has more clarity than one would expect for concepts they haven't previously had access to. One possible sign that this kind of memory overwriting has been occurring is a memory from childhood that has an quantum shift that has an extremely clear and precise memory without explanation.

The possibility of weak memories merging by this means is increased since memories are rebuilt every time they're visited. The more a memory becomes stronger, the more frequently it's used since the brain has more experience in reconstructing the idea. This is why it's important to study and why one can become experts in a particular area after years of refining ideas on the same topic. It is logical to conclude that the reverse is applicable, and the memory is weaker when it's not significant enough to be used frequently. This implies that it isn't an accident that Mandela results to become associated more frequently with ideas that aren't regularly reconstructed. The blending of ideas could be a possibility if the brain believes the mix of these lesser frequented memories is feasible enough. Reconstructing a memory which has been merged in recent times due to being discussed as an Mandela change can create the illusion of trust in the newly created memory.

The notion that memories are blurring because of wear and tear lends itself to countermeasures employed by the brain to fill the gaps in memory with plausible substitutes. This generally non-intrusive method for dealing with memory gaps is done by using schemas and stereotypes. Memory is like a puzzle that has been recreated and the brain prefers to complete a part of the puzzle using an unintentionally inaccurate information instead of having gaps. Furthermore, the existence of plausible replacements for certain aspects in memories reduces the stress on the brain's ability to store accurate information. This compromise of accuracy of a memory and preservation of brainpower is particularly beneficial for gaps that are connected with a minor aspect. It makes sense for individuals to experience identical gaps within their memory due to the fact that the memories that are associated with Mandela influences are more prone to being brittle. The schema and the stereotypes that we learn in our society are alike enough to allow an entire group of people to fill in the gaps in memory with the same possible replacement.

A number of flaws that are inherent in the use of frequently accessed memories, makes it possible to believe Mandela effects to result from confabulation. Confabulation occurs when someone's memory is replaced with an assertion that they continue to believe to be correct. The trust in the accuracy of a memory could persist even when the evidence is clear that the person's memory is inaccurate. The brain's natural tendency to internal consistency may be a hindrance to the person who is experiencing confusion when incorrect memories get reinforced by individuals who have the same idea.

The repeated use of false information from multiple sources helps the brain keep the falsehood in place. The mental gymnastics that a person is able to perform to keep internal consistency over false information is amazing, however, it is also a bit unfortunate. The possibility of confusion could explain the reason why groups of people recall the exact Mandela changes similarly. It's possible that the necessity of maintaining the sameness of these false memories is sufficient to ignore the contradicting evidence which is apparent in our current reality.

Memories have been contradicting the reality because of the misinformation effect for a long time prior to it was even known that Mandela effect was discovered. It is reasonable to attribute a mix of variables, including the possibility of memory being misattributed and suggestibility in the development and spread of quantum shifts. The effect of misinformation has been shown to affect people in critical situations, for example, when they testify in the courtroom. Lawyers should avoid asking questions that are leading because it could affect the memory witnesses are accessing.

Information that is trivial (such as movie titles and music lyrics) are likely to have a greater chance of changes because of the possibility of suggestibility and misattribution of memories than information accessed through witness testimony. Additionally, inaccurate information that is responsible for the misinformation effect could occur at any time during the process of deliberately or inadvertently disseminating lies. The repeated repetition of false information and reinforcement by confirmation bias can help clarify the way to have memories that relate to Mandela changes to be able to attain an extremely high degree of certainty.

It is crucial to recognize that individuals can simultaneously believe that their memory is flawed, and that Mandela effects aren't always due to misremembering certain details. My memory, for instance, is clearly imperfect, which made me to see the value of misremembering as a reason to explain Mandela changes. I would continue to believe in quantum shifts being the result of poor memories if I hadn't come across instances that I couldn't think of as rational.

The lack of explanation consistency led me to dig deeper into the subject and discover ways to explain these cases or to explore alternative explanations. In the event that rationalization failed in some of the examples I looked at and it opened the way to take the notion of an alternative explanation seriously even if it turned in being unprovable. It is likely that some theories related on the Mandela effect may be caused by factors like misrememberance or cognitive dissonance. misperceptions. If one instance cannot be explained by these kinds of explanations, it may open up interesting possibilities.

The variety of ways to miss-remember or miscommunicate the information is so numerous that every alternate memory that is attributed to an Mandela change is recalled the same way. This unusual lack of variation is unique given that every person has at the very least different data in deciding on the best replacement for weak memories or gaps. This issue can be prevented if an alternate version of the same element is an occurrence of widespread confusion.

It is likely that there will be at least some variation in the event of the communication is not clear, due to the fact that information shared is likely to alter due to subsequent instances of communication issues.

An excellent example is the game of telephones, that is designed to show how easy it is for miscommunication to occur and the variety of different variations that can occur in the span of a few minutes. If the miscommunication and/or memory issue causes Mandela effects It makes sense to have multiple groups be present for every Mandela change , based on the variant of the issue they recall.

Another aspect that supports the possibility of an alternative explanation is that it is a relatively new phenomenon, even though humans have been able to recall their memories for many thousands of years. The fact is that we today have technology that offers unparalleled methods of connectivity and storage of information. However, the availability of new methods of communication, like the internet, is not enough to explain the suddenness in this technological advancement. Actually, they were in existence for a long time prior to even the Mandela effect was ever discovered.

Conceptions about the mind have been in the news for a long time therefore, a more alarming kind of discord between memory and reality is likely to remain by unnoticed for many thousands of years. A misremembering of the collective at the level of Mandela changes is likely to be an old and well acknowledged phenomenon if poor memory is at the root of this problem. This isn't only the emergence of the concept that's novel, given the recent discoveries of the term in the course of researching the troubling mismatches between their memories and the reality.

A variety of universes that are parallel is a popular explanation for the appearance of Mandela changes. The power of explanation for this possibility is due to the alleged possibility of having multiple parallel universes which are just a little different from one another. In the event that coexisting universes is a possibility this would help clarify how shifting consciousness between two universes will only cause small differences.

The book will concentrate on the technique used to create universes that rely on quantum mechanics as it's great for explaining parallel universes that differ a little. Details that are slightly different in each reality clearly explains the way that some individuals can be affected by quantum changes and how others can retain the information precisely. People who remember the details accurately are most likely to be to come from a universe in which the details are the same or are from their own universe. The idea of a multiverse, where the consciousness of a person can shift between two universes may be an entirely new concept and that is the reason why it's a new idea.

The existence of residuals is a concern no matter if the appearance from Mandela changes is due to the inability of memories or the mobility of consciousness within the multiverse. Mandela result residue can be described as a different representation of an aspect , such as a meme or parody. Residue is significant because it reflects the alternative version of the aspect that seems to have changed. An individual who remembers an aspect in the form of residue may use it to prove that something has changed.

However, residue may be used to illustrate the vast scope of misremembrance, or to explain how the concept behind the alternative version was conceived. The presence of residue is not a good explanation for the Mandela effect of misremembrance as it proves that an alternative explanation is possible. Remaining with residue that is similar to the representation of an element from the person's original parallel universe exceeds the limits of a plausible coincidence. This is due to the sheer number of possible differences in a particular aspect is enormous.

The way parallel universes with similar properties are created is by waves breaking down in every continuously branching universe. The collapse of an underlying wave function can be described as the moment at which wave functions cease to be in a superposition state. Particles typically are in superposition and this is a contradictory situation that involves attributes like location , which is represented by probabilistic results instead of a fixed value. The particle encompasses all possible outcomes in the same moment that it is in superposition. The process of expanding universes includes the possible results of a collapsed wave function occurring with each result merging into a different parallel universe. There are a lot of very similar universes, as this is the way the universes form due to the fact that the collapse of just one wave function is not a significant factor. This staggering amount of parallel universes is the reason those with a shift in consciousness are able to have different memories as the majority of details correspond to their initial universe.

The inexplicably conclusive conclusion that particles are in a superposition state has been repeatedly proven by the double slit experiment. A single particle that is directed towards the double slit shows the probabilistic behavior of making an interference pattern on the measurement device that is placed between each of the slits. The particle that is the only one is able to generate an interference pattern since the potential paths covers both slits.

It is obvious that the possibility of probabilistic results for each slit that occur simultaneously causes the possible outcomes of each slit to conflict with one another. This wouldn't be possible in the event that each particle didn't behave as a wave with possible locations, instead of having certain characteristics. The ability to simultaneously cover each possible outcome prior to when measurement is made isn't unique to single particles. The interference pattern has been observed when a single atom molecules are used in double slit tests.

It is evident that the reason why a particle is not in superposition for a long time is due to the capability to gauge different properties of each particle. A different possibility is that of an event of collapse that occurs when a particle is in contact with the environment in a way that can cause disintegration. This alternative explanation for the collapse of wave functions is not plausible given that the fact that the particles of the superposition of a star stay throughout billions of years, as demonstrated by their continuous emission of light. The fusion reactions inside the star responsible for emitting light are dependent on the particles remaining in superposition . This means that quantum tunneling is feasible.

Fusion reactions wouldn't be possible if particles already had a set with certain characteristics, rather than being in a low probability of achieving quantum tunneling. If the constant interaction of an astronomical star for a period hundreds of thousands of years aren't enough to trigger decoherence then it's unlikely that to be the case that this alternative explanation for collapse of the wave function is more convincing than the concept of measurement being the primary the determining.

An example close to home that also illustrates the deficiency of explanations for decoherence are the interactions in photosynthesis. The first stage of photosynthesis is an electromagnetic exchange between electrons and photons that are found in chlorophyll. This leads to a transfer of energy to an electron. It is a maze-like route the excited electron must travel within the space of a nanosecond to ensure that energy is stored in a safe manner in the plant's reaction center. It's likely that the only reason that photosynthesis can occur with almost the complete efficiency of photosynthesis, in spite of the complex maze the excited electron has to navigate within a remarkably short duration, is because it remains in superposition.

The energy packet could not get to its correct location if the electron's interaction with the photon or the interactions on its way towards the reaction center could be sufficient to collapse the wave function. If the chaotic environment inside the leaf isn't enough that wave functions break down and collapse, then the concept of decoherence is no longer a valid explanation significance for the data gathered in the double slit study.

The quantum eraser experiment with delayed-choice is a further piece of the puzzle to determine the feasibility of branching universes using wave-function collapses. This experiment adds additional features to the double slit test including a crystal to create entangled pairs of particles, several forms of measurement, as well as the removal of the results of certain tests. One of the particles is examined on an unlit screen, without any attempt to analyze the what-path data. That means all particles will display the pattern of interference.

The outcome of the particle that is measured on the screen will be determined by the reaction to its twin that has delayed measurements of its what-path information. These particles are measured at two distinct times in time, and the measurement of the data from which path occurs when the display has taken measurements. The fascinating result of this study is that the pattern of a particle appears on screen based upon the data from its entangled twin is observed.

The main factor that makes the various results displayed on the screen useful in determining the feasibility branches in universes their capability to intentionally loss of the which-path data in the delay measurement. One of the most fascinating outcomes is the appearance of an interference pattern in the event that the which-path information for the measurement delay isn't recorded, while an interference pattern doesn't show up on the screen when the which-path information of the delayed measurements is captured. If the initial measurement and/or interaction with the screen determined the ultimate outcome, all the results would be an interference pattern since the wave-function of its entangled twin would be collapsed prior to the time when the delayed measurement occurred.

If the measurement delay of the particle that was entangled was sufficient to make the interference pattern disappear on the screen the pattern should not be a problem if the outcome of the delay measurement is recorded. The deliberate deletion of which-path data in the delayed measuring (resulting to an interference pattern appearing on the screen) indicates that the reduction of an oscillation function is contingent on the outcome being recorded.

The multiverse's branching can cause an issue with timing, when the mechanism is the same process. It is crucial to try and figure out when the many results of the multiverse have become feasible outcomes for our universe as they already have branches into their respective universes. The issue lies in the timing of branching when parallel universes (by the use of wave function) fall apart if branching takes place prior to a measurement taking place.

Wave functions' existence would be a mystery If all possible outcomes represented by the function were branched into isolated universes prior to any interference occurred. If the branching takes place in the course of an interaction or measurement, then multiple outcomes cannot be observed using the delay-choice quantum eraser since the branching will occur prior to the delay measurement taking place. If two measurements or interactions are sufficient to branch of universes, the different results could not be seen when the monitor was turned on, contingent on whether the data is stored or deleted. This could mean it is possible that the splitting of universes may be inconsistent or is a result of a mechanism that has more complexity.

Chapter 2: Mandela Substitutions

There are at most three types of quantum fluctuations. One of them which will be discussed within this section includes Mandela substitutions. Substitutions can be described as the smallest of changes, like words being spelled in a different way or a letter in the word being replaced or words in a film or song becoming distinct. Arguments over the quantum nature of these changes are hard to establish from both sides of the argument between people who believe the change happened since the present and alternative versions are highly interchangeable.

These are great examples to illustrate the possibility of memory playing part in the creation of Mandela changes as we have a tendency to misremember this kind of information. People who argue from the perspective that the change is happening is generally less able to deal with substitutions , given how widespread mistakes and mispellings are. It's still a crucial class to consider as they represent the most prevalent kind of Mandela effect that can be seen to arise. Following Mandela substitutions are just a tiny portion of the total amount of examples available.

Forrest Gump is a memorable film that was awarded the audience six Academy Awards, three Golden Globe Awards and six Saturn Awards, and a number of other prizes. One of the most famous lines from Tom Hank's character Forrest Gump is "My mom always told me that life is like a chocolate box. You don't know what you're going to receive." Another version many people have in mind can be "Life can be like a chocolate box. You don't know what you're going to receive." The use of"was" as a word "was" included in the quotation is logical in explaining the confusion due to misrememberance considering that Forrest could be quoting what his mother used to tell him.

The fact that the world remains similar to a box of chocolates is most likely not understood by the character due to his lack of intelligence. There could be some evidence to the previous existence of the quote that had "is" in place of "was" and the misquoting of the famous quote on the boxes for the DVD and VHS. It's a bit surprising that this error wasn't corrected on the DVD box , even although it came out several years after the VHS copy.

Oscar Mayer is a company that is well-known for bologna and hot dogs as well as other types of meat. Their spelling Oscar Mayer should be familiar to a lot of people due to their catchy tunes that prompt customers to shout the company's name and even spell the name. There's a gap between those who remember the name currently used by Oscar Mayer and those who are aware of the company's previous name, Oscar Meyer. One reason for those who are not remembering correctly can be the sound of the letter "a" Mayer not being the long vowel sound.

There's an 'e' in second part of its name therefore it could be a possible substitute if one's recall of the company's name is not strong. One possible source of evidence to support the alternative version of the name of the company is a category that was featured on the show Jeopardy that aired on the 27th of April, 2005 which used"Oscar Meyer" "Oscar Meyer". It's interesting that none of those who created the category, or in the associated questions were able to be aware that the company's name was written incorrectly.

Let's Call the The Whole Thing Off is a song from 1937 in which there are comparisons made in between British American and British American English accents. The lyrics of the song utilize "like" to make comparisons, whereas others use "say". One of the lyrics that has endured through the years is "You love potato, and I love potahto. You like tomatoes and I love tomahto". Another memory associated with this line is "say" being used in place rather than "like". People who remember "say" in place of "like" could be mentally changing the lyrics, and it is more logical. "Like" is a word that "like" can make it appear like someone was offered alternatives and chose the one they liked, but this isn't the usual situation.

The pronunciations are usually learned and most people stick with the pronunciation they are comfortable with. There's a definite contradiction in the lyrics when looking at the consistency of pronunciations, and also the variations of pronunciations used in singing "like" in place of "say".

Berenstain Bears Berenstain Bears are well-known characters who have appeared in novels as well as a TV series as well as video games. The popularity of the Berenstain Bears hasn't kept many people from having a different perception of the characters being called Berenstein Bears. Berenstein Bears. It's natural that people misremember that the character's name is Berenstein instead of a variety of possibilities because "-stein" is more often employed as an ending to names. There are a variety of differences in the memory of people. are more likely to occur in the case of the Berenstain Bears due to the fact that a large portion of people haven't been exposed to the name prior to their childhood.

A possible reason to this change is in the way the name was incorrectly pronounced in a commercial on TV in 1994, which advertised the Sega Club. Nobody was able to fix this error during the creation of the commercial, even when there's no doubt that he used "Berenstein" in place of "Berenstain". The error that made its way into final versions of the advertisement could be more appropriate when the characters were not popular at the period.

Snow White is a classic fairy story that Disney turned the story into an animation film, in 1937. One of the most famous quotes from the film can be found in "Magic Mirror on the wall Who is the fairest among all?" A group of people will remember the evil queen's remark "Mirror Mirror at the back of the room, who's the fairest among all?" There are many ways that this quote could have been linked in the movie, for instance the phrase used of the classic Brothers Grimm tale being "Mirror mirror, mirror".

The line "Mirror mirror" is so often used to refer to the film of 1937 that it's often misquoted on merchandise that is Disney-branded. Reenactments and parodies of this scene often misquote the phrase as "Mirror reflection on the wall". The line is incorrectly quoted throughout Shrek, The Fresh Prince of Bel Air, The Simpsons as well as other films and television shows. The sheer amount of times the line appears to be misquoted is odd enough to raise the question as to whether the quote from the movie's original script has been altered in any way.

Even the Bible does not elude obvious modifications. Different interpretations of Isaiah 11:6 speak of the wolf who lives with lambs and a lamb, which is an excellent metaphor for a serene period to come. An error that is often made is when a wolf is substituted for an lion in the verse. The verse that has the wolf in place of the Lion makes sense considering that wolves are predators on sheep. The present version serves as an image for a tranquil period because a shepherd will not be required when the wolf is living with the lamb , rather than trying to consume the animal. But it is true that the Bible is one of the most popular books to read, learn from and make imagery for, making it harder to imagine that there are so many people portraying the verse in a wrong way. It is normal for individuals to make this error but the vast majority of images available for this verse depicts an lion, not the one.

"Objects in the mirror are more distant that they look" is a standard safety message on the side mirrors. It has been repeatedly viewed in the eyes of millions over time. A large portion of people have a different version which states "may may" in place of "are". The message shifts from certainty to an uncertain state this is unusual since the intent and/or message are usually retained to a greater extent in the alternative version.

It's possible that people might be confusing this warning with other warnings which say "may be the result". A fascinating example of a possible reason to this possibility of change is David Letterman's top ten list from February 13th 2001. The top 10 list of things that sound cool when said of James Earl Jones included "Objects in the mirror might have a closer appearance than the ones they show" at the top of the list at number 9. The inclusion on the list's top 10 may indicate a shift or simply a reminder of how prevalent this possibility of confusion is.

Scary Movie is known for its humorous parodies of many different films. The only exception is the copy-paste of a well-known phrase from the film The Sixth Sense. "I am seeing dead persons" is the line used in Scary Movie and is even featured on the film's poster. A few people have recalled the phrase to mean "I am seeing white folks" on the poster, trailer and the film. It is possible that the false memory be a result of Undercover Brother, which is comedy in which a character laughs when he says, "I see white people".

The argument to change the version is how inappropriate the current version. The use of "I have seen white persons" in the film would have created a humorous scene that would be in line with the other hilarious scenes in the film. "I have seen dead bodies" is also not in line with the overall tone of the film. The fact that it appears on the film's poster is not logical since it makes it appear that the film is openly taking inspiration from other films, instead of its true intention of parodying horror films.

Scary Movie 2 parodies supernatural films such as The Haunting, Poltergeist, The Exorcist, and The Rocky Horror Picture Show. There's a hilarious scene where the character Dwight Hartman would rather fall to death rather than play Hanson's small hand, which has small fingers. Dwight requests Hanson's other hand however, Hanson responds, "My other hand isn't strong enough. I'll take your little hand". A few people will remember

Hanson declaring, "Take my strong hand". An untrue memory could be the cause of this alternate interpretation because Hanson speaks of the smaller of his hands as the "strong hand" earlier in the film.

Also, there is a noticeable lack of consistency throughout the film, as the character's voice abruptly shifts between referring to hand strength to dimensions of his hands. There are clear signs in the film that point to the character's ignorance about his left hand's size being different, therefore removing the schtick this way is puzzling. The possibility of a recurrence is also present in memes that say "take your strong hand" in place of "little hands".

The Lone Ranger is a fictional mask-wearing hero who has been featured in a TV show, radio show as well as in books and films. "Hi-yo, Silver, away!" is the Lone Ranger's catchy slogan and has been well-known to the public over the last few decades. The catchy phrase is a surprise to many who remember the phrase to mean "Hi-ho, Silver, away!". It's possible that the thoughts of the "Lone" Rangers catchy slogan are mixed with the memories of "Heigh-Ho" that was sung by the seven dwarfs from Snow White.

A minor distinction between the latest version and the alternate memory may cause confusion about which one is the right one. It is quite certain that this alternate memory was used in the scene with the projector of Ace Ventura: When Nature Calls and in Bruce Almighty, the Saleen S7 scene in Bruce Almighty in Bruce Almighty, and The Lone Ranger from 2013. The alternate version that appears in a film that is inspired by the character with similar name particularly awkward.

Kellogg's Froot Loops is a widely well-known brand of cereal available in many countries and has been in existence since 1963. The alternative source of the name for this delicious breakfast cereal would be Fruit Loops. It is natural that people associate the name with Fruit Loops since it's fruit-flavored. It's also natural to have to use the two "o" within both of the words because it incorporates the product it self into the logo as a substitute for vowels. A reference to the alternative version is possible in the Van Halen 1982 World Tour rider, which famously prohibits Brown M&Ms from the band's dressing rooms. Fruit loops is mentioned among the possible options for breakfast cereals that are served in the 8:30 a.m. hour a.m. every day of a concert as described on pages 36 and 37 of the riders. The rider is very meticulous, so it's quite shocking that a spelling error was omitted from an official version.

Star Wars: Episode V Star Wars: Episode V The Empire Strikes Back has one of of the most well-known and shocking statements in cinema history. It is known as "Luke Your father is me." The real meaning of the quote is "No I am your dad." The shocking nature of this revelation could be a factor in people possibly misremembering this line. It's also sensible to substitute "No" by "Luke" when reciting it to someone else as it makes the message simple to understand. One of the most convincing pieces of evidence is found as James Earl Jones's distinct voice, which is a variation of this phrase. It can be heard by pressing the Hallmark ornament of Darth Vader from 2005. If this is real residue that is awe-inspiring, since this audio recording from the alternative line didn't alter while the current variant of the lines can be present in every version of the film.

The peanut butter brand is a typical staple in the kitchen, but there are some notable brands, including Jif, Skippy, and Peter Pan. But, some people still remember the top Peanut Butter brand Jiffy. Many people can even recall the shift between Jiffy into Jiff before changing to Jif. This is a change that takes place in multiple steps instead of three distinct varieties of peanut butter, since individuals can easily recall all three versions in different dates. The names of peanut butter brands could be trivial enough to allow the brain to blend the Skippy peanut butter and Jif peanut butter to create Jiffy peanut butter. Of the many possibilities of leftovers to come up with an alternative name for these peanut butters, definitions of the term "jiffy" from the Urban Dictionary won out. Two definitions are referring to peanut butter in the definition of the word. This is fascinating because the website uses different groups of people to publish definitions, and yet they all missed the error of connecting peanut butter in two different definitions of the word.

One popular show that many have watched throughout the decades has been Looney Tunes. This may not be the right choice for people who are part of the people who remember the show's name as Looney Toons. It's not a stretch of imagination to imagine that people are unable to remember this because of the show being an animated show, not an actual music performance. The terms "tune" or "toon" being the same does not help for being able to remember the correct spelling. An odd disconnect can be observed when you compare characters from two different cartoons that are part of the same universe of cartoons. Tiny Toon Adventures uses a distinct spelling from its predecessor, even though the characters from the show are heading to college to be an upcoming generation of Looney Tunes. It's more logical to blame it on misrememberance If Tiny Toon Adventures was a copycat of the original however that's not the situation.

Field of Dreams is a popular film which was nominated for the three Academy Awards. One of the most memorable scenes involves Ray Kinsella hearing a voice whispering, "If you build it and he comes, he will come." The most well-known alternative to the phrase is "If you construct it and they come, it will be built." The audience could easily forget this particular aspect of the quotation because the usage of "he" isn't making any sense within relation to the film. People who are confused by this quote could have mistakenly rearranged this contradiction to the movie by substituting "he" by "they".

There are multiple players in the crowd and the crowd comprises more than one person which can further cement the wrong version of the phrase into people's minds. The possibility of a residue can be found in a similar situation in Wayne's World 2, in which the protagonist is told that "If you create it then they will come".

Sally Field won an Academy Award for Best Actress in a Leading Role for her performance as a lead actress in Places in the Heart. "I must admit that you love me. At the moment, you're liking my personality!" is a memorable line from her Oscar acceptance speech in 1985. Many people can recall an alternate version of the line in which she said, "You like me, you really like me!"

Sally Field says, "Let's look at which one you prefer. You really love it!" when she announced the Best Actor during the 1986 Oscars. Both of these quotes were made in the same context which is why it makes sense to have the memories of both quotations to be combined into the alternative version. One could argue it is possible that her quote of 1986 could be an oblique reference to the alternative version of her acceptance speech. Another piece of evidence can be heard when "You are my love, you truly have a love for you!" is used as an imitation acceptance speech in the film The Mask.

Mister Rogers' Neighborhood is one of the longest-running children's TV series . It has a well famous opening theme "Won't I Be Your Neighbor?" Mister Rogers started every episode by walking in the front doors and singing "It's a gorgeous morning in the community." A number of viewers remember Mister Rogers being more inclusive through his singing "It's an amazing morning in this neighborhood." The false memories may be the reason, given the influence of childhood memories as well as other versions with "the" in place of "this" like the theme song from the show Daniel Tiger's Neighborhood.

A Beautiful Day in the Neighborhood is a film inspired by actual events from Fred Roger's life. This causes the title's change of the title a bit confusing. A simple error within the movie's title can be hard to overlook since everyone involved in the process of making should be aware of the movie's name. It could be a quick and simple fix if a one of them noticed the error and pointed out the error.

The Silence of the Lambs is a psychological horror film that was released in 1991 that has been hailed for being one of the top movies of the year. Anthony Hopkins plays the cannibalistic serial killer Hannibal Lecter and Jodie Foster plays a student at the FBI Academy known as Clarice Starling. Hannibal Lecter famously said, "Hello, Clarice" when Clarice was interviewing his serial killer on the first occasion. The issue with this is that it was actually a quiet "Good Morning" in lieu of "Hello, Clarice".

Hannibal Lecter isn't the only one to use this line until the trailer for Hannibal that was released in the year 2001. It could be an opportunity to misremember the exact words spoken at their initial meeting. This doesn't make sense of parodies that popped up in between the release of the two films , such as Ernie "Chip" Douglas saying the phrase during The Cable Guy, which was released in 1996. It is evident which source this parody comes from. Jim

Carrey's character's words "Silence of the Lambs. Hello, Clarice. It's nice to see you once again."

The significance of using the residual from television and movies is related to the amount of people who are involved in these kinds of projects. A lot of examples of residue carry more uncertainty due to the fact that they are based on only one person's memory or a task completed by a small number of people. The validity of attribution of false memories to an alternative version is dependent on the number of participants.

The possibility of false memories being associated with a fragment of residue will decrease as the number of people who are exposed to this part of the process grows. If the present version of an feature existed prior to when the original scene was conceived or performed and later edited, then it has a good chance of being rectified. It's outside the realm of acceptable coincidences for each participant in a massive project to have exactly the identical false memory of the same aspect. Thus, the possibility to use Mandela effects is enhanced because of the persistence of residue in films as well as other projects that require large numbers of people.

Chapter 3: Mandela Shifts

The second kind of quantum shift is a Mandela shift that encompasses various possible elements that go beyond the realm of the words changing. The rationalization of false memories that accompany the more complex changes could be problematic since the possibilities are greater for doubt. Not being able to remember a single word from an iconic film in the same way as other people can be understood since the possibilities of replacements are dependent on the circumstances of the line and general scene.

Recalling the same information for the more complicated aspects of Mandela shifts is fascinating because there are a variety of possibilities of combinations that could be involved in the creation of an incorrect memory. It isn't possible for a shift to occur due to multiple people inadvertently forming an alternate version of the same thing and/or propagating an incorrect explanation in a way that creates an entire group of people that has the exact same false memory. The power balance between those who claim to have misremembered the event and those who believe that a change is taking place is on an than equal terms in the case of Mandela shifts.

Video footage from President John F. Kennedy's assassination, which occurred on Nov. 22, 1963 has been studied and watched by a variety of people throughout the decades. The most recent footage of this historic event clearly shows a number of six people inside the car. Many people are aware of an alternative version with only four seats were in the car. The viewers tend to be focused on the assassination attempt of JFK which is why it might be very easy to miss or overlook the additional set of people within the car. The possibility of JFK shooting in a vehicle with only four seating is from the one that is on display in the Historic Auto Attractions in Roscoe, Illinois. Other museum exhibits utilize models of the car that has six seats. However, nobody who were involved in the exhibit in Historic Auto Attractions observed this error. The inconsistent cars that are used in museum exhibits creates some uncertainty as to whether there has been a change.

A lot in the cast of Alice in Wonderland are memorable particularly Tweedledum or Tweedledee. The 1951 film adaptation depicts these characters wearing curvaceous torsos, clothes of mixed colors as well as the red hat that has tiny flags sticking out of the top. A few people have recalled that both characters looked identical, with only their hats that were multicolored and propellers sticking out the top. The caps that are worn of Tweedledum or Tweedledee are kind of similar to propeller hats. the flags spin as propellers.

This, along with the fact that they have a poor memories of their caps may cause memories of flags being replaced with propellers. It is natural for people to mistake that the hats are multicolored, since other parts of the clothing incorporates a variety of shades. The common memory of propellers in multicolored hats is a long-standing tradition that has allowed this mix of elements to be present in different costumes and representations. However, these costumes could be the result of a transformation as well as the representation of Tweedledee and Tweedledum seen within The Simpsons season 28 episode 11.

The See-Threepio (C-3PO) can be described as an inscrutable protocol droid that appears in a variety of Star Wars novels, television shows, comics as well as the Star Wars movie trilogies. One of the things that distinguish his droids from other protocol droids is the silver coloration of his right leg's lower part and foot. This can be a little troubling to those who remember his lower part as being a monochromatic area. This part of C3PO's physique is usually hidden and is difficult to discern because of background colors or even in dim lighting. The combination of these factors and the memory of gold color throughout his body can result in an illusion of feet and legs as being monochromatic.

The absence of silver on the leg of C-3PO is evident in comic books and action figures, as well as in stickers, as well as the animated television show Star Wars: Droids: The Adventures of R2-D2 as well as C-3PO. It's odd to a show that is centered on the adventures of C-3PO to ignore this aspect, especially if the color of his right leg isn't changed.

E.T. The Extra-Terrestrial is a well-known movie about E.T. being trapped on Earth and revolves around a bond he develops with an infant named Elliott. The scene in which Elliott is stranded shows E.T. is learning new words to tell the kids that they are the other aliens. The message is conveyed with the phrase, "E.T. home phone" Then both children and E.T. saying, "E.T. call home". Some people remember E.T. always saying "E.T. home phone" instead of saying "E.T. home phone" first time.

A false memory in this line is likely to be a plausible form in light of the fact the fact that "E.T. Home phone" is only said once, while "E.T. call home" is mentioned in six instances in the scene. There is also an intriguing scene in the film's debut trailer, where "E.T. call home" is sung rather than "E.T. Home phone". Then we hear "E.T. telephone home" as E.T.'s shadow appears. E.T.'s hand is visible on Elliott's face. This echoes the memories of E.T. saying the same thing over and over again. Uncle Sam is an iconic symbol for his country's United States' federal government and his present appearance dates in time to World War I. The initial model of the hat had an white crown with a large blue band, with large, white five-pointed stars. Another memory associated with the hat is that it has vertical red stripes across the crown, giving the hat a strong link with that of the American flag. The blue band of white stars in the present version is similar to that of the American flag, which means that people could easily include red stripes in the memories they have of Uncle Sam's cap.

The potential repercussions for the alternative model of Uncle Sam's cap is overwhelming in its quantity. For instance the alternate version was utilized in the movie Apollo Creed during his entrance to the title match in Rocky. The original versions are harder to locate than the alternative version which is unusual for Mandela modifications. There's a plausible explanation for the large amount of residue for this unique phenomenon that is discussed in chapter 5.

The Thinker is an internationally known statue created of Auguste Rodin that has been utilized to symbolize the philosophy of the mind and has been recreated many times. The first version of this change that was that was associated with the statue was the statue's fist shifting across its face to below its chin. It's normal for memories to mix over the location of the statue's hand given the fact that both poses depict the person who is thinking. However, what doesn't make sense is that people confuse the pose while they're replicating the statue's posture as they stand next to The Thinker.

This is precisely what can be observed in photos of people or several people place their hands on their foreheads although the statue appears to have an entirely different posture just next to them. It is normal for people to be conscious of the correct posture when in the presence of an statue, and the photographer does not see a difference.

The image of the statue that shows him with his fist on his forehead, and the other one with his fist under the chin is both wrong. The latest model that is part of The Thinker clearly shows the statue's hand underneath its chin, with an unlocked hand. The differences in an unopened hand beneath the jaw and a fist is subtle enough to justify the possibility of the emergence in false memories.

What's interesting about the Thoughter is its possibility it's gone through two distinct Mandela modifications. The people who were photographed have managed to get the head and chin misplaced and also fail to recognize the distinction between the fist and open hand. There is also the difference between the memories that have become distorted due to many years of not being used and the new memory of those who visited and taking pictures in front of the statue. The Thinker could be an important part of the puzzle as the statue appears to have changed several times and the remnants of photographs didn't change accordingly.

Tutankhamun known as "King Tut" is well-known as the youngest pharaoh in Egypt and also for the artifacts like his gold mask, which was discovered in his well-preserved tomb. Ancient Egypt has become a sought-after research topic as King Tut's gold-colored mask has become a popular artifact used to depict the rich Egyptian history. King Tut's gold mask is one of the most famous death masks with Tutankhamun's resemblance and features an imposing vulture and Cobra in the face. A few people have recalled that the mask was similar to the one seen on the original, but it having a cobra that is lone instead of two animals like it appears on the real version. It's possible that the cobra is more famous to people, which could explain why some people seem to have forgot about the vulture that was that is on the mask. One of the possible pieces of evidence is the costume used by Tut in the King Tut in the appropriately known television miniseries titled Tut. The cobra can be used as a visual signal to the viewer that is why the absence of a vulture a point to be noted.

Pikachu is among the most well-known species of the Pokemon franchise because it is the main mascot and also appearing in a variety of videos, games and animated shows. Pikachu is predominantly yellow, with an orange circle on each cheek , and 2 brown stripes across its rear. Also, it has an area of brown at the bottom of its lightning bolt tail, and the tops of its ears appear black. A few people recall the mouse-like Pokemon having a black part at the tail's end instead of having an orange section at the end of it's tail. It is possible that this memory mistake could occur if people imagine adding black at the end of the tail in order to match their ears' black tips, or integrate its black tail from Pichu to the look of Pikachu. A possible relic of this alteration can be seen within The Simpsons season 24 episode 4. In where Maggie's character features Pikachu's distinctive black-tipped tail and ears , while also having an underlying brown color to the tail.

The Mona Lisa is a masterpiece created by Leonardo da Vinci that is one of the most important well-known, most well-known and visited paintings around the globe. The most recent version depicts a woman with a smile and curly locks sitting upright in an armchair, in front of a fictional landscape. The alternative version is thought of as having a vague smile that is not as obvious as the smile that is seen on the latest version. The persistent doubt over The Mona Lisa's smile might result in memories being shaped by this notion, which could result in people thinking of the smile as being more unclear as it really is.

The other version is shown in the picture taken from The Da Vinci Code with Tom Hanks and Audrey Tautou facing the painting and mimicking Mona Lisa's absence of a smile that is obvious. It's odd that the reproduction was able to replicate the painting so accurately, except one minor deviation from one of the aspects that make the painting so popular.

The Statue of Liberty is a famous torch-wielding copper statue that is based upon Liberty, the Roman goddess of liberty, which France gave its citizens to the United States in 1886. The statue is an iconic New York City landmark and an official national monument situated in Liberty Island and is visited by millions of people every year. Some individuals have an alternate view of where they believe that the Statue of Liberty had been in Ellis Island. The alternative version of the statue's location may be a result of the more than 12 million immigrants who were processed at the immigration station located on Ellis Island. This amount of immigrants connects the concept that liberty is tied and Ellis Island even though the Statue of Liberty is located on an island. A possible remnant of the statue's location situated on Ellis Island is found on the 1986 Statue of Liberty commemorative silver dollar, as the coin reads Ellis Island on it. This is rather odd since the coin clearly is commemorating the statue, not the idea of liberty.

Ed McMahon is well-known for being a co-host in The Tonight Show Starring Johnny Carson presenter of Star Search, and appearing in films and commercials. McMahon was the initial representative of American Family Publishers, which offered subscriptions to magazines and also held sweepstakes with huge sums of cash. Some people recall Ed McMahon working as a spokesperson for a competitor firm called Publishers Clearing House. It is also believed that he was thought of as a frequent visitor to the house of the winner of the sweepstakes, with a large cheque, and this is a bit odd as he did not perform the role of a spokesperson as a spokesperson for American Family Publishers. It's not a stretch of imagination to confuse a popular spokesperson for sweepstakes with a well-known business that offers sweepstakes. A possible case study is an episode of Late Night with David Letterman where Johnny Carson presents an oversized check to Letterman. Carson refers to Ed McMahon repeatedly in connection to the check. It clearly states Publishers Clearing House on the check.

Poppin Fresh, also known as The Pillsbury Doughboy is widely known as the Pillsbury mascot. He has appeared in over sixty commercials over the many years. The iconic character is known for his dough-like physique as well as his chef's hat with white neckerchief, blue eyes and the giggle that he makes when he is poked in the stomach. The distinctive difference in the appearance of the Pillsbury Doughboy that is observed by a large group of people is centered on his neckerchief being blue, not white. A possible memory that may be a reference to the neckerchief's color can be found in the blue sailor's neck worn around the neck of Stay-Puft Marshmallow Man from Ghostbusters. Possible evidence of this is The Pillsbury Doughboy having a blue neckerchief, while the Mr. Burns hallucinates in The Simpsons episode 7 of season 7. Poppin Fresh is also seen with a neckerchief of blue as being smashed to death by Lois during Family Guy season 1 episode 7. It's fascinating that the crews involved in the two shows seem to have misinterpreted this famous character similarly.

Cruella De Vil a flamboyant villain who has a desire to slaughter dalmatian puppies in order to get the feline coat with a spotted pattern. Cruella from the Disney's 1961 version of 101 Dalmatians has a distinctive and memorable appearance because of the bony physique, black, white hair and long red gloves and her large, cream-colored fur coat. A different version where Cruella De Vil's cream-colored fur coat gets replaced with the dalmatian fur coat, is revered by a crowd of people. The simpleness of Cruella's fur coat, combined with the strong connection with dalmatians may be a reason for an untrue memory of her sporting a dalmatian-fur coat. The black and white hair may contribute to the mixing of memories that include Cruella's spotty coat and the Dalmatians. One interesting aspect to think about is the sheer number of costumes of Cruella de Vil which include spots that resemble dalmatian, but do not accurately represent the appearance of this character.

The various versions of the coat can have an influence on the story because Cruella De Vil's motive is to design an dalmatian-style coat. From a first glance it's logical to Cruella De Vil not to wear a spotted coat since it serves as a rational reason to her motivation to create a coat with a pattern. But, the alternative version is more logical because she's easily established as a skilled villain who has probably slayed Dalmatians prior to. The movies are designed for kids, and so the over-the-top way in dressing Cruella De Vil with a spotted coat is not a bad idea. Adults will easily understand why she is aiming for the dogs to wear the sake of a spotted coat. The motivation of Cruella could be overlooked by children as the simple coat doesn't reveal her motive for having puppies. The absence of spots in the latest version of the film is more likely to reduce the tension in the film, and possibly confusion young viewers as to what is the reason why the villainous lady needs the Dalmatians.

Tony Tiger Tiger is the well-known feline mascot of Kellogg's Frosted Flake. He first came into the public eye in 1952, and has seen several design changes throughout the time. One characteristic of this cartoon mascot that seems to have never changed throughout the years is the shade that his face appears to have. Tony's nose is always blue, even though a number of people are united in their memories that his nose was black. The nose color of Tony's is a simple thing for people to forget because the color of a cartoon's nose of a tiger will not be a major factor in their lives. Family Guy season 2 episode 14 depicts Tony the Tiger wearing a black hat in the company of breakfast cereal Mascots in a parody to The Breakfast Club. Family Guy season 5 episode 11 is a parody of Kellogg's mascot dubbed Terry The Tiger. It is interesting to note that Terry the Tiger is shown with an orange nose, and Tony The Tiger appears depicted wearing a black nose.

The Heisman Trophy is an annual prize in the field of college football to the player who is the most outstanding as judged by their outstanding ability determination, dedication, perseverance and commitment to work. The prestigious trophy depicts players who use the stiff-arm fend while having both feet on the ground. This model has been in use since the first time it was awarded in 1935. Some people are aware of an alternate design of the trophy in which the athlete has one leg elevated into the air. The alternate pose is simpler to do and the higher number of people who lift their legs for the position could lead to people recalling the posture of the trophy incorrectly. Another interesting thing to note is that people are getting photographed in the alternate posture while carrying the original trophy, for instance Barack Obama when he was the Democratic presidential candidate. Tattoos of Micky Mouse performing his Heisman posture to Hines Ward's upper right hand is a different part of the residue, since it shows the alternative model of the position.

The United States of America is a country that is highly developed within North America that is known for its freedom in the economy, the quality of life, as well as military expenditures. The country is home to fifty states and a federal district. There are also five major territories that are not incorporated. There are a lot of people who recall of the United States having fifty-two states although they aren't sure as to what the additional two states were. A contributing factor to the possibility of confusion is the distinction between the states as well as federal districts getting blurred because of Washington, D.C. having three votes in the Electoral College. The lines are also blurred between territories that are not states due to the fact that people who were born within Puerto Rico, Guam, Guam, the U.S. Virgin Islands, and the Northern Mariana Islands are considered citizens of the United States. The potential evidence for the existence of 52 states is within Brewster's Millions when the amount of states are mentioned . two lawyers discuss the scope of Monty's ads for the mayoral elections.

Human hearts are a vital muscle organ that circulates blood to circulate nutrients and oxygen. This vital part of the human body is situated in the middle of the chest. Many people are aware of the fact that the heart is located on the left of the chest , as demonstrated by those who are instructed on the wrong hand placement when reciting the Pledge of Allegiance of the United States. The heart's beat is evident more on the left side since it houses the largest pumping chamber in the heart that can be responsible for triggering circulation of blood. The more apparent heartbeat on the left side may possibly contribute to the notion of the heart being situated on the left side of your chest. This possibility of change could be an element of relic that relates to the biology instructor on Dead Like Me season 2 episode 9 specifically telling her students that the heart isn't situated in the middle within the body.

The Karate Kid is a 1984 film where an adolescent boy named Daniel LaRusso learns karate and life lessons from an Okinawan immigrant named Mr. Miyagi. The most recent version of the film features Daniel LaRusso wearing an iconic white headband that has the blue lotus design. The headband appears to have changed from a headband of white with the red design of the rising sun in accordance with the shared memory. The shared memory may come in the shape of the circle that appears on the national flag of Japan in addition to Japan being known as"The Land of the Rising Sun. "Land of the Rising Sun." This also is logical considering the fact the fact that Okinawan Karate is a factor in the martial arts depicted within the movie. A possible relic is seen in the movie the film Kickin' It Old Skool In the film, Justin Schumacher is tricked into an amusing montage of chores following his recovery of the unconsciousness. Justin is wearing a bandana that features the red sun rising and is particularly comparing the concept of recuperation through the use of chores to Daniel's training.

Traffic lights are lights used to regulate the flow of traffic on roads and are seen by millions of people every day. Stoplights are the same color code, that indicates red lights. Red light blocks traffic from moving while yellow indicates there's an imminent red light and green indicates that traffic is proceeding normal. The arrangement in which the lights are displayed is in line with red being the highest with yellow at the middle and green on the bottom. Traffic lights with an upper green and red in middle and red on the bottom is popular among people. One possible explanation for the shared memories of a different design is the green light that is over the red light used for signals on railroads. The Simpsons episode 17 from season 17 depicts a traffic light in an alternate configuration in an episode in which Marge regrets her absence of companions. This is a bizarre mistake to make when you consider the uniformity of the current configuration.

Charles Lindbergh Jr is commonly called "the Lindbergh baby" due to his kidnapping and murder at the age of 20 in 1932. The tragic incident led to being called the "trial in the year 1900" that is a bit odd because there is a popular belief that the Lindbergh baby not being found. If memory is at fault then the widespread understanding of the Lindbergh baby's kidnapping was somehow disassociated from the infamous trial that followed the death of his father. The gap between the details of the kidnapping as well as the trial isn't impossible considering the length of time since the incident occurred. There are many sources and examples of the alternate memory , including an episode of

The Simpsons season 7 episode 8 where Grampa Simpson informs the FBI that he's an actual Lindbergh baby. Sam & Max Beyond Time and Space is an online game where two police officers who are freelance who are from the modern era discover Charles Lindbergh to be a talking child because of his young-age fountain.

Wendy's is currently the third largest fast-food hamburger chain, with more than 6,700 restaurants and is well-known for its hamburger patties that are square as well as the Frosty and the the use of the freshest ground beef. The company is famous for its corporate logo with the daughter of the founder, wearing her straight, red hair in hair pigtails. The corporate logo appears to have changed based on some people who remembered the girl's braided hair that was pigtail-style. The alternate interpretation is caused by misrememberance, by mixing Wendy's mascot and the look of Pippi Longstocking, who wears braided hair that is sticking out in a sideways fashion from her head. It's also simple to confuse braids and pigtails because they are both a good match and don't differ much in appearance when not paying to. One of the most compelling evidence of possible residue can be evident in an Wendy's commercial that ran in 2007. entitled Hole in which the main model wears a red haired braided wig that is clearly visible.

Zero was Jack Skellington's dog from The Nightmare Before Christmas and accompanying his owner on a few of his adventures. Zero is an eerie pumpkin-colored nose which perfectly matches with the overall theme Halloween the land. The bright, glowing nose of Zero is handy when guiding the skeleton and sleigh reindeer through the hazy skies on Christmas Eve. The glowing pumpkin nose that is visible throughout the movie will be a pleasant surprise for those who have a memory of his nose being red and not sporting an oblong face. This is a plausible memory as the distinction between orange and red isn't as obvious in the context of a dog's nose.

Additionally, the use in Zero as a substitute for Rudolf could lead to confusion of colors. Incredibly, Zero is shown as sporting an red nose in the hardcover edition of The Nightmare Before Christmas that was published by Golden Books and Disney. It's harder to pin this error in the production of the VHS tapes since that the book came out in 2021.

The Jungle Book is a Disney film from 1967 , in which a little boy called Mowgli is adopted by wolves, and then befriended by other animals , such as the sloth bear called Baloo. The most memorable scenes of the film is Baloo sneaking into a group monkeys in order to save Mowgli by disguised wearing a grass skirt lip shaped coconuts. Many people will remember King Louie being fooled into believing that Baloo was a female wearing an ointment made of coconut instead of the coconut pieces used to disguise lips. Both costumes are equally ridiculous which is why it's a bit silly to build a real argument on how convincing a cartoon costume is. It's possible that conflation resulted from grass skirts and coconut bras that are worn in Hawaii and the surroundings of these islands being comparable to the settings of the film. One interesting bit of evidence is Baloo from the Disney film Once Upon a Dream Parade wearing a coconut bra, even although he didn't wear it in the film.

Chapter 4: Mandela Subtractions

The third type of quantum change is called a Mandela subtraction. It is distinguished by the absence of a distinct element such as a phrase, word, or other entity. The fact that this kind of change is astonishment considering they retain the distinctness and vivid memories that accompany the two other kinds of Mandela effects. The examples for subtractions aren't as likely to be caused by a cluster of people who fill in gaps in memory since there shouldn't be a space to fill the memory aspect that was never remembered existed.

It can be difficult to reconcile two perspectives in the event that one aspect appears to be nonexistent in the reality. It is therefore harder for false memory to be given the benefit of doubt when it comes to Mandela subtractions. Finding evidence of a feature which has never been seen can provide more compelling evidence of an eventual change. The power balance between those who believe that there was a misrememberance and those who believe that there was a change heavily favored by those who believe in a change. Therefore, they must consider subtractions in proving the validity of the claims.

Star Trek: The Original Series is a science-fiction TV show that features William Shatner as Captain James T. Kirk. A single of the famous phrases used in the character of Captain Kirk was recited and recalled in the form of 'Beam me Up, Scotty!' However the line was not utilized in any episode of the show or the films. It could be because of confusion and/or misquoting other phrases like "Scotty beam us up", "Scotty, beam me up" and "beam me up".

It could be a matter of the incorrect version being repeated over and over again that people do not know "Beam my head up Scotty!" wasn't an expression that Captain Kirk. The huge popularity of this quote could also be the reason James Doohan, who played Scotty utilized it as the title for his autobiography. It's not common to have an actor perpetuate the most popular quote associated with his character , and then name his autobiography this way which is why it could be the result of a shift.

Richard Simmons is known for his workout routines, in which his energetic and lively personality blends with upbeat music to provide a relaxed 1980s-style training sessions. One of the most prominent features that a lot of people recall as being associated in Richard Simmons is a headband that he wore regularly during his workouts. But, the headband is recalled by many people is not present on any of his many photos or videos. The rumor that Richard Simmons wearing a headband isn't a huge leap given his workout routines that cause sweat, like the appropriately named videos titled"Sweatin' and the Oldies.

It's fascinating to observe how effective the distinctive headband was in expressing his character through jokes and costumes, since he did not wear it. One parody can be found on the rock star's Modern Life episode titled "No Pain no gain" where

Richard Simmons voices an aerobics instructor. This parody is fascinating since they managed to make this mistake while involved with the individual they're parodying.

Fruit of the Loom is an American firm that manufactures clothing , including T-shirts as well as underwear. The company's logo is a well acknowledged logo that features red apples and purple grapes, as well as green gooseberries, grapes, as well as green leaves. You may be wondering what the reason for the absence of the cornucopia from the description of the logo If you're among those who have a memory of a certain variant of this logo that been missing for a long time. It's possible that this alternative memory is a conflating of the assortment of fruits in the logo with the array of food items that spill out of cornucopias during the seasons of Autumn. The logo for underwear seen during South Park season 16 episode 6 could be a bit of leftovers since it states "Cornucopia Label" and includes an apple and grapes placed on top of the cornucopia. This is a fascinating choice because the fruit is likely to remind viewers to the Fruit of the Loom while using a cornucopia as a part of the name and the image doesn't hit the mark.

Sharing experiences with memories of the Fruit of the Loom logo with a cornucopia offer an important angle to consider for this phenomenon. The shared belief that a cornucopia is an loom because of limited vocabulary skills is a crucial viewpoint to think about when evaluating the validity of certain Mandela modifications resulted from false memories. On the other side, the typical outcome of people mistakingly believing that a cornucopia is an loom because of common context clues can be positive reinforcement for those who are who think and store memories similarly.

Fruit, for instance, is used in the logo and is mentioned in the name of the brand It is therefore logical to associate the cornucopia to the last word in the name of the brand. On the other hand this shared experience of relating images and words can be challenging from the point of misrememberance because the cornucopia wasn't featured in the logo. The people who reached the same idea using the cornucopia a source regardless of the fact that none ever actually saw a crop when they made the link in the event that a change hadn't occurred.

Moonraker comes as the 11th installment within the James Bond franchise with Richard Kiel reprising his role as an ominous henchman, with strong steel teeth. The character played by Richard Kiel is known as "Jaws" as the film is notable for the scene in which the character of Blanche Ravalec is called "Dolly" as they are in love. Some people share a common memory of Dolly looks at Jaws and then reveals the braces she wears on her teeth. The memory is not in line with reality since she doesn't wear braces. Dolly wearing braces when they are in love with each other is plausible given the amount of metal that is on Jaw's teeth therefore a gap in memory could have been filled by several people.

The possibility of a residue could be an Finnish commercial for Sampo Mini Visa Card in where an episode from Moonraker is a parody. Richard Kiel is in the commercial, and he offers the cashier a huge smile, followed by her smiling and showing her braces.

Jaws along with Dolly getting married is an odd scene to be included in the film considering Dolly doesn't have braces. In addition, it's uncommon for a villain to include the beginning of their romance explored during the movie's running time however, the film's makers clearly believed it was worth the effort to include. It's likely that the directors chose to show the moment of the couple falling in love since it showed how even characters similar to Jaws is able to find people who have similar traits to them.

This message is clearly not received when Dolly doesn't have braces since their most similar feature is absent and this causes the scene to feel unnatural. A convincing relationship builder isn't the forte for Bond films, making the presence of a scene demonstrating this kind of relationship more bizarre. The film could have shown Jaws as having already one, and nothing could have been either lost or gained, since there is no clear reason for their instant attraction.

Lassie is a TV show that was on for 17 seasons and is well-known for the main character, the Rough Collie dog as well as the boy Timmy Martin, voiced by Jon Provost. One of the most well-known episodes of the show focuses on Lassie seeking assistance for Timmy after Timmy fell into a well. It is among the most well-known scenes of the show's history , even although Timmy did not fall into the well in any episode. It was Lassie who plunged into a well in the two-part episode of season 17 titled "For the love of Lassie".

It's possible that there was a false memory regarding Timmy being swept into the well, given the number of other locations he fell into like lakes, rivers as well as quicksand and an abandoned mine shaft. It's incredible how many references and parodies are available for this famous story that have never been seen. The most intriguing sources of possible residual is the memoir The Timmy's In the Well Jon Provost's Story. Jon Provost Story.

I Love Lucy is a TV sitcom that first ran from 1951 until 1957, and was the most watched television show on the United States for more than half of the period. The show centers on the lives of Lucy and Ricky Ricardo and humorously follows Lucy's unsuccessful attempts to enter show business. Ricky declaring, "Lucy! You've got some explaining to get through!" is one of the most well-known catchy phrases from the show, even although he did not say it on any show. The possible false memory could be due to Ricky using variations of the catchy catchy phrase in a similar manner to the way people recall the original version employed. The popular catchy catchy catchy that seems to never have been used in the show has been a hit countless times, while it is clearly absent. Fools Rush In shows one the most likely instances of residual in which Alex Whitman uses the seemingly missing line, rather than any of the variants that are in real world.

Kellogg's Raisin Bran is a breakfast cereal with the slogan "two spoons of raisins per box" and also has an animated sun that is named "Sunny" to serve as the symbol. Sunny supports the slogan of the product by putting two scoops of raisins on top of an empty bowl of Raisin Bran. Sunny appears to be lacking a unique feature, as per those with an enduring memory of wearing sunglasses. Many people remember that the sunglasses played a part in an advertisement that featured sunlight as the key ingredient to make the raisins taste great. The most likely cause of confusion could be glasses used on The California Raisins that were very famous at the time. People remember vividly Sunny with sunglasses on in ads. Parody version of Sunny is seen dropping huge raisins from two scoops during Family Guy season 1 episode 2. One aspect of this parody that doesn't seem to make sense in light of our actual reality is the addition of glasses.

The adaptation from 1995 to Pride and Prejudice is a British television show starring Colin Firth as Mr. Darcy and Jennifer Ehle as Elizabeth Bennet. The original show was watched by more than 10 million viewers and received the most positive reviews from reviewers. The scene in which Darcy. Darcy is seen walking across a lake with his shirt hanging over his chest was thought to be among the most memorable scenes in British TV history. The scene ended up being among the top TV moments , even although it didn't happen as per all evidence from the show. There are no moments of. Darcy entering the lake swimming, walking, and then leaving the lake. His heartfelt departure from the water is clearly absent. It's possible that the people filled in the gap in the Mr. Darcy exiting the lake by stitching other scenes. Oddly enough, several films and shows have paid tributes to the scene, even although it is not evident that it ever existed.

The Twilight Zone is an television series of anthology developed by Rod Serling, which has been a huge hit due to the many kinds of genres that are involved as well as the unexpected twists. Many people will remember several, if not all, episodes from the show that started in Rod Serling saying, "Imagine that you could imagine!" ..." The line is intrinsically tied with Rod Serling and The Twilight Zone even though he hasn't began a single episode using the line. The repercussions are absolutely shocking in the alleged existence of this famous phrase, which includes memes, articles claiming to be referring to the phrase as a reference to the phrase, as well as Futurama parodying it on The Scary Door referencing the inexistent line.

Rod Serling and The Twilight Zone are often named when an individual or article borrows this helpful introduction. The degree of belief that a change occurred in this particular line comes down to the likelihood that so many would be able to be able to identify the source of the term in a wrong way. The thing that is more bizarre is there is no evidence to support the possibility of confusion of this particular phrase.

Monopoly is a well-known board game that revolves around purchasing properties and destroying opponents with rent and trading. The mascot of the game is an old and wealthy man wearing a mustache of white, a morning suit and a black top hat. He is often called "Rich Uncle Pennybags", "Mr. Monopoly" or "The Monopoly Man". The most memorable thing that appears not to have been included in Uncle Pennybags is the presence of a monocle that covers his eye, according to a common memory. One possible source is the monocle used by the mascot from Planters named Mr. Peanut since they each have a cane and wear the top hat. Ace Ventura: Pet Detective offers a fascinating fragment of possible evidence in the context of a scene Jim Carrey's character's voice says the man in his 80s "And you're"the Monopoly person!" This comparison is fascinating because the man wears an elongated monocle, however he does not have the other distinctive features of Rich Uncle Pennybags, such as a top hat and a cane.

"Curious" George is a chimpanzee popular for the trouble he causes during his adventures , which are chronicled in a collection of books as well as a television show and even a film. The husband and wife team comprised Margret Rey as the author as well as Hans Augusto "H. A." Rey as the illustrator. The choice not to include a tail indicates that he's an ocelot, but also the subject on this Mandela subtraction, as a lot of people have a memory of the tail he had.

George is often referred to as a monkey many times throughout the story of the character It is therefore not a stretch to refer to George as a monkey, instead of an ape. Additionally, the distinction between chimpanzee and monkey can be fuzzy for some because "monkey" is often a general word used to describe a variety of primates. These elements could result in people recollecting Curious George with a tail as most kinds of monkeys have tails. thought to have a tail.

Curious George was introduced as a character shortly before the beginning of World War II while Margret and H. A. Rey resided in France. Curious George was in fact named "Zozo" within the United Kingdom around this time that was probably to prevent him from the use of the name King George VI's name to refer to an animal. The change from George's name is beneficial in the sense of finding possible remnants of his tail. H.A. Rey created a set of cartoons starring "Zozo" in Good Housekeeping Magazine, in which the cartoonist clearly had the appearance of a tail. It's not surprising that to use the title "Zozo" was chosen since this magazine also appeared in the United Kingdom at this point. The comic strips, which were six panels long featured a variety of examples of George having fun, like setting off a firework inside an office. The comic strip that was featured used "Zozo" for the title and showed Curious George moving around quickly on the record player. This particular instance is a colored pencil drawing by H.A. Rey that was published in the 1951 issue of Good Housekeeping Magazine.

Risky Business is a well-known film from 1983, starring Tom Cruise in his breakout role as Joel Goodson. A first act Joel did to entertain the house was to dance in his shirt and underwear as the track "Old time rock and roll" played in the background. A memorable element of the scene is not present due to the common impression that of Tom Cruise wearing Ray-Ban sunglasses as he slid across the floor, spinning around. Some people mistakenly believe that Cruise was wearing Ray-Bans in the famous dance scene isn't out of possibility considering that they appear on the movie's poster as well as are worn by Joel in the other scenes of the film. There are many recreations of the dance scene which include sunglasses, such as The Goldbergs season 3 episode 1. There isn't much doubt as to where the idea was for Barry Goldberg's efforts to glide through the air in underwear, given that the show was heavily in the style of Risky Business.

Shazaam is a movie in the mid-90s in which Sinbad portrays an enchantress. The problem is it's an untrue memory shared by everyone since Sinbad affirms that he was not a part of the cast in the film and the film appears to never have existed. One of the most troubling aspects that this memory shared is that the plot seemed to be so general and unrememberable that it is difficult to remember a lot of specifics. The non-memorable plot and dialogue are difficult to recreate because these memories stem of people who watched the film several decades ago. It is believed that the Sinbad the Sailor movie marathon in 1994 was hosted by none other than Sinbad dressed as a genie, which may be a source of residue or confabulation. On the other side, he might have dressed as a genie when hosting the marathon to promote his film Shazaam. On the other hand Sinbad wearing the costume of a genie at the time of the marathon could be a crucial element in the creation of shared memories which merged to create a film called Shazaam.

Sinbad was producing films at the time and the alleged film Shazaam is a film that he could have appeared in. It is believed that people do not confuse Shazaam with Kazaam which was the 1996 genie film that stars Shaquille O'Neal. Many people vividly recall Kazaam as a clear copycat from Shazaam even though the former was never made. A fascinating piece of remnants can be evident on All That season 2 episode 13 where Sinbad's character says "I I am Sinboo." The phrase is similar to Sinbad declaring "I I am Shazaam" In Shazaam as well as Shaq declaring "I Am Kazaam" within Kazaam.

The first film did not exist while the sequel was made its debut a few months after this episode was aired. Another possibility of a hint at the film's absence is Sinboo being in arms positions that resemble other genies like the one from Kazaam. The smoke used to show that Sinboo was introduced to the room may also be an indication of Sinbad's genie emerging from the lamp.

Britney Spears, a performer, singer as well as a songwriter and dancer and is one of the most successful artists, with more than 100 million albums sold around the world. Her album Oops!... I Did It Again peaked at the top of the US Billboard 200 and sold more than 1.3 million copies within the first week. "Oops!... I Did It Again" was the first single from the album. It also had its own music video which Britney dressed in a memorable Red PVC bodysuit. The microphone headset that many recall her wearing in this music video is now gone. One plausible explanation for this possible misunderstanding is the fact that Britney Spears was wearing a mic headset in other occasions, such as live performances. Lizzie McGuire season 1 episode 2 could be a possible source for the headset in memory because Hilary Duff's character visualizes her self as Britney Spears in"Oops I Did It Again "Oops !... It's Happened Again" music video. Did Them again" music video. Lizzie McGuire wore the remembered headset, as well as the red suit. Her animated alter-ego wore a headphone with no red dress.

The Matrix is a science action thriller made in 1999. It explores the notion of people who do not realize they are living in a virtual world and features Keanu Reeves playing Neo along with Laurence Fishburne as Morpheus. When Neo encounters Morpheus for the first time, certain people will are able to recall Morpheus telling Neo, "What if I told you?" ..." before you confirmed that the Matrix really exists. It's highly unlikely that it was an exaggeration because nothing has replaced the line that seems to be missing as well as other scenes from the movie do not resemble each other. The movie was released before memes began to become famous, and it's not unimaginable for the people who made their own Morpheus videos in the year 2012 to make use of an expression that was never actually stated.

The creators of these memes taking liberties with their creativity could have created the false impression about

Morpheus saying the phrase. This doesn't provide a reason for why Morpheus states, "What if I were to say that you're not experiencing the world as it really could be" In South Park season 11 episode 11. The fact that this episode aired in 2007 suggests that it could be a relic from change, as it predates the first Morpheus memes by four years.

The line that I remember of The Matrix apparently disappearing is the reason I realized there could be something happening beyond the concoction. I thought about whether I could have made up the line I was able to vividly recall in my mind after having watched the film so many times. If I could have invented it, then what would have made so many other people create the identical line in their own memories? I would have accepted a false memory in the event that the line was slight altered, but I would not imagine the fact that many people are misremembering an iconic line even if it never existed.

I spent the next few years pondering what kind of parallel universe, time travel or a simulation is the best way to provide the best explanation for this Mandela effect. The idea of the idea that time travel was associated went out of the window since residuals are impossible in this scenario. The concept that there were parallel universes was longer to dispel from a personal perspective and was achieved through numerous thought experiments based on the double slit hypothesis.

The absence of these options left open the possibility of being a fake or an overwhelming number of coincidences that revolve around the exact same small details. The concept of coincidences appeared ridiculous given the exactitude involved in people misremembering exactly the same elements in the exact same way , even the fact that some of these features appear to never have existed. There was no way that a bad memory could be able to disprove this line disappearing or any other Mandela subtractions, and the time came to contemplate the other explanations that could be the most plausible.

I also realized my limits of what is considered a credible coincidence, after having written an essay on deliberate evolution. The notion of Mandela subtractions as a random coincidence could be more plausible when it was a singular event. But even a single event that is convincing can call the legitimacy of misrememberance to the fore. It's only fitting that a line missing in The Matrix is what led me to thinking that we exist in a computer simulation.

Chapter 5: Simulation Theory

The explanation I offer the possibility that the development of Mandela shifts, substitutions and subtractions is based on our experience as a society as part of a hyper realistic simulation. This is the most convincing reason for the existence residual since a realistic virtual reality could have flaws that be noticed by those who are in the simulation. The advantage of residual over other possible inconsistencies within a simulation is its apparent newness of its appearance even though the changes are propagated retroactively.

This allows us to concentrate on the relatively recent innovations that could lead to the appearance of Mandela effects. Other potential issues within simulations would be very difficult to analyze in the same way. This chapter will focus on the reasons the reason why Mandela changes may occur in the simulation, rather than other issues like the reason the simulation even exists. The following suggested limitations of the virtual reality scenario might help explain the reason this phenomenon began to emerge quite recently.

It can be difficult for people to tell if they are having an inaccurate memory, as demonstrated in numerous studies. However, it is possible to recognize an event as false memory if evidence is discovered that does not support the memory that was influenced. It is much more difficult to discern the memory that was written over in a simulation because reality is usually altered to correspond with the new memory. If the memories of a few individuals were correctly overwritten to reflect the new reality, while other people retained the memories of the original aspect the same as the Mandela effect might occur. People with an improperly overwritten memory might decide to ignore this small error because they are aware that their own memory isn't flawless. The second group may also decide to believe that a shift is taking place and formulate theories about how reality has changed. The members of the last group who are looking for reasons to have alternate memories are rightly considered odd since they believe in memories that aren't in line with reality.

The development the emergence of Mandela changes is due to the notion of the simulation running on a device which is not connected to an extensive amount of resources. It is crucial to think about since a large number of resources will be required to load each particle in the universe, in the case of an actual simulation. It is logical to minimize the requirements of resources using techniques for efficiency like illusions, if resources are not sufficient. If the system is able to access infinite resources, it's very unlikely for significant discrepancies resulted from overstressing the system to appear in the simulation.

Thus, my concept of the system's inaccessibility to resources that allows glitches to happen would be impossible if the limitation does not exist. These limitations also presume that the system operates upon pragmatic principles and the system is provided with sufficient resources to allow the simulation to run according to its it is intended. If there is a limit in the amount of resources, there may result in consequences if the limits are exceeded.

One sign that we live in a virtual world that is governed by an intelligent system would be the constant presence of superposition even when the measurement recorded hasn't been observed. Variables represented as the possibility of a variety of outcomes is likely to be less energy-intensive than keeping defined states for every particle. Utilizing approximations of variables, instead of having defined characteristics is advantageous as the former is something like a question mark and the latter needs specific characteristics to be preserved in storage, altered, and stored in the event of changes.

There is no debate regarding whether superposition is real, as the double slit experiment shows that this state is a real-world reality. A crucial question is whether superposition could be a reality in a computer system with access to an abundance of resources like computing power. Superposition in an simulation would make much sense in the event of an insufficient amount of resources to loading the massive quantity of data our universe could require.

The disintegration of wave functions is an effective last resort to convince people that the real world is not a figment of their imagination. The constant existence of superposition have never had negative consequences across millions of years. This implies that a particle's capacity to cease being in superposition once the measurement is recorded is unique in the overall world of.

For instance particles interact with each in accordance with physical laws regardless of whether the wave function of their particles has been collapsed or not. The only difference between superposition and the collapse of wave functions is the data that is gathered through the recording of a measurement. This would be a massive alarm if the recorded measurements didn't produce defined data, as it indicates an absence of clearly defined quanta at the quantum scale. Thus, the advantage of allowing particles an established state will likely be greater than the advantages of maintaining particles in superposition. The quantum level function similarly to the macro level whenever an event is recorded helps to maintain the illusion that particles are identified objects, instead of showing their probability of existence.

The fact that the collapse of wave functions is a problem due to the fact that they appear to have no real use in the universe. Furthermore, it appears that there is only one option to get a fully collapsed wave function, and that is by recording a measurement. The possibility of having collapsed wave functions is logical when it was necessary for the functioning of the universe. If particles required decoherence in order to effectively interact with each other for creation of galaxies the recorded measurements had to be taken at the beginning of our universe. If measurements recorded were not needed to create the universe so why are wave functions acting differently in relation to when the universe was formed? This is the kind of inconsistency that occurs the case if decoherence occurred in the past with no recorded measurements, whereas this requirement now is the reason for the breakdown of the wave function. The universe appears to have been capable of functioning correctly throughout billions of years, while particles were constantly in superposition. This suggests a deficiency of explanation power for the capability to create an actual collapsed wave function.

The power of explanation for the existence of multiple universes due to collapsing waves loses their force if superposition has been the norm for the universe for thousands of years. If multiple outcomes of an event of collapsed wave functions aren't possible without a documented measurement and the mechanism that is responsible for making parallel universes has been in operation for a brief period of time. This could drastically decrease the number of parallel universes because experiments taking observations at the quantum scale may be the only method by which to create this kind of collapse.

This is an issue because it transforms the creation of universes that are parallel from becoming commonplace to being unimaginable up until a few years ago. The power of explanation for parallel universes by referring to the Mandela effect could be diminished because certain changes appear to have had a retroactive effect on the past prior to the first observations made at the quantum scale. If the branching phenomenon wasn't feasible at a particular date in time, it is highly unlikely that parallel universes were involved in changes that occurred prior to.

Parallel universes are branching regardless of whether a wave function actually collapsed, then a theory dependent on this kind of collapse of a wave function will lose all explanation power. The type of branching described above would be a result of every wave function occurring in a distinct universe at any moment without disrupting superposition. There must be an explanation of a distinct difference between a true collapsed wave function and the continuous wave function collapses which don't disturb superposition.

In other words, there is no explanation for the diverse outcomes that are observed in the double slit test since there is nothing different in the scenarios. This is due to the fact that each branching scenario could occur regardless of whether an recorded measurement was made or not. The existence of a gap between the two possibilities implies that the branching model will not be able to explain anything with regard to the physics of it and Mandela changes. Another possibility is to explain the reason particles stay in superposition after a wave function is collapsed instead of each particle retaining the same outcome that they are able to

104

define in line with the new branched parallel universes.

The simulation's use of resources can be reduced further if superposition is not sufficient for reaching the desired efficiency levels. One approach is to project realistic illusions to other areas of the universe, so that even particles in superposition have to be kept in superposition. This is a great method of conserving resources as we are not able to physically examining these distant sections of the universe.

The creation of this kind of illusion is possible by using the holographic technology. A precise illusion that can be achieved at large distances will result in drastically reducing costs for resources, and does not compromise an immersion into the model. Our models of cosmic events work without taking into account all particles that could be involved, therefore it is possible for sophisticated recreations of regions of the universe in a hyper-realistic simulation. The model with more techniques of efficiency could be able to explain the development of Mandela changes since it suggests that the system has limitations that could be overcome.

One way that the system could utilize more resources than it intended is by generating data in recorded measurements. The generation of this previously undefined data requires resources at a greater rate over a long time. It is also possible that there will be an increase in the use of resources to maintenance of the previously generated data. The drain on resources diminishes once the information has been created and the data is now in maintenance mode. The amount of resources used up would correspond to the volume of information generated during the time.

For example the collapse of the wave function as a result of taking the information about the how-way of one particle is likely to have less impact on the system than measuring simultaneously the properties of several particles. The resource drain could be reduced by borrowing resources from other processes in the simulation until the data has entered maintenance mode. This may be a challenge with regard to the resources available for other processes in the simulation, as a lot of data is created through recordings of measurements.

CERN (the European Organization for Nuclear Research) is renowned for its work with The Large Hadron Collider (LHC) to record the results of the collision of charged particles. This kind of experiment on a quantum scale will likely result in the simulation using more resources when the results of the recorded measurements are generated simultaneously. The volume of impacts could cause the simulation to be overwhelmed when triggers are not utilized to limit the amount of information gathered.

The appearance of Mandela effects is also coincident with the schedule of operation at the LHC. For instance, Fiona Broome attended the conference , where she met people who remembered Mandela's passing in the 1980s . This was the year in which the LHC began to perform collisions. However, the nature of retroactive of quantum fluctuations can make it difficult to discern when a change actually took place. It's not as difficult to determine the relationship or the coincidence of spikes that occurred that are associated with discovering Mandela changes in relation to run dates for the LHC. There is some kind of connection or positive relationship between the quality in Mandela effects that appear in the data gathered by CERN's particle experiments.

The importance of resource limits and the stress on the system due to the generation of data is the reason for problems that can occur. The possibility of glitches that alter or erase parts of the simulation will normally be avoided if adequate resources are in place.

One time that there may not be enough resources to maintain the corrective measures is when resources are transferred to other functions within the system, for instance when the precise information needed for particles are generated. It shouldn't be a surprise to find that significant amounts of the resources diverted to other tasks in LHC experiments, given that approximately 140 Terabytes of data are produced in just one period of time. The obvious spikes in the appearance of Mandela changes over the course of tests conducted using the LHC is a good indication of this kind of resource diverting. This is a significant indication because these kinds of spikes shouldn't be observed if the alternate memories are the result of confabulation. Instead there must be a clear absence of any correlation with things like date of run of the LHC when fake memories were involved since these could appear sporadic.

The original version's residue could be due to glitches that cause changes in the simulation aren't able to distinguish abstract copies of the same aspect. This is particularly true in instances when the copy is employed in inventive ways that aren't explicitly connected to original material, such as parodies and memes. So, the accuracy of glitches in this simulation are directly related to the amount of similarity in the content of original and the duplicate of this feature.

For instance someone who is standing close to The Thinker as they perform the pose originally isn't likely to be recognized through the process of overwriting. The person isn't with the statue when performing the same pose. However, it's difficult to identify the source of Mandela substitutions since the modifications are too minor for humans to discern a duplicate enough to allow the glitch to be overlooked. Finding duplicates of an aspect that are so obscure that they do not warrant the suspicion of glitches is an effective method of determining between people who are experiencing an Mandela alteration and one with a false memory.

The complexity of interconnected memories and the lack of resources or overwriting accuracy may be responsible for the difference in memories that are of the current version and the alternative version. This could explain why each one is affected with an unknown range of

Mandela changes. This is more complicated by the possible capacity of those suffering from cognitive dissonance due to malfunctions in memory to alter their memories in an attempts to eliminate the inconsistencies that exist in their memory.

On the other one hand, this can happen if someone isn't able to accept the thought of an aspect changing , and their brain rewrites instances of alternate memories to shield the person from this contradiction. However it is possible for people to believe in their own memory so that their brains erase instances of conflicts in memory even if these alternate memories don't correspond to the reality. This effort to ensure consistency within the brain could be similar to the way that our brains fill in gaps in memory by generating plausible substitutes instead of leaving gaps. People who share memories of the present or alternate version of the same thing is an interesting topic that requires more study to understand more fully.

It is possible that there is an increased importance in the overwriting of the original version of an aspect , based on how close one is to the part which has been altered. One explanation could be that cognitive dissonance is more intense in the case of a direct link.

For instance, if the organ is relocated or locations, a person with backgrounds in medicine will feel a greater sense of cognitive dissonance due to the alteration. The increased feeling of cognitive dissonance could significantly increase the chance of memories relating to the original version being replaced by memories associated with the present version. This could also explain how people in South Africa were not shocked by Nelson Mandela becoming president in 1994 because they didn't have shared memories of his passing. If Mandela passed away in the early 1980s, before an Mandela change was made and a change in the presidency was made, there was a need to erase the memory of his death of those living in South Africa.

The insignificant aspect of Mandela effects is probably the reason why these modifications haven't been manually corrected. There are likely to be modifications that we haven't heard of that were changed manually because their impact was too obvious. In other words, it's an unintentional coincidence that the majority of Mandela modifications have had minor aspects. Some of the more notable changes could be worthwhile pausing and/or rewinding the simulation until a significant change is occurring and the consequences that accompany the change is fixed.

However, small changes that could be rationalized through confabulation aren't as important to fix manually. It is likely to not be worth the time and money to fix any aspect that isn't significant enough to cause a disruption inside the simulator. But, it is an extremely risky option since a plethora of tiny modifications could cause the same disturbance as an enormous change that eventually causes immersion to be broken in the simulation.

Chapter 6: Quantum Mechanics

One of the things we need to look at in detail are quantum mechanics, and quantum physics.

To make sense of this, I'll provide a very basic explanation. The simplest way to define an understanding of the term "atom." The word"atom" is in Greek literally means not to be able to break it into pieces smaller or be indivisible. It's inevitably incapable of being divided.

The idea is the Greek philosophical idea of the smallest unit of matter that could never be separated. Atom is the unit that makes up the elements, and the elements that are combined make up molecules.

A very specific arrangement of an atom will provide you with the element. The molecular structure of objects is more than just one molecular structure. It are the foundation of everything we have that is both living and non-living. It becomes bigger and more massive as it organizes into larger systems.

In modern times, we've discovered that atoms can be split and we can have sub-atomic particles. Everyone, I'm sure has heard of the electron, and later a proton and neutron. These particles are the source of particle physics. In general, logic would suggest that if you've got one and cut it in half, you must cut it into cut it into halves. If you choose one of the halves, and cut it in half, you can continue doing this until you reach the age of infants or until you have no more.

It's fascinating to note that once it reaches the quantum realm is that you'll find"the Planck Scale. It's the smallest and most small-sized baseline of matter. On the Planck Scale, nothing can be smaller.

To provide you with an understanding about what I'm trying describe in a brief period of time. The length of the Planck is calculated by 1.616199 using 97 zeros in the following it is multiplied by 10 to the negative 35th Power in the metrics. This is what we call the Planck Scale.

On the Planck Level when they cut subatomic particles and particles and keep breaking the particles by smaller or smaller pieces, once they reach the Planck Level, the particle is co-located or co-relatively situated in layman's terms. This signifies that it is in multiple places at the same time. This is the beginning of a new realm known as quantum Entanglement.

To comprehend this, we'll make use of one experiment. I'll discuss it quickly in a straightforward manner. A scientist called Young in the year 1801, carried out an experiment known as the double slit test or the Young's Slits or Young's Experiment. What it does is generally uses the light source. It shines it onto plates and there's a screen on the plate. the experiment, there's a opening in the plates which will let light pass through it onto the screen. In another experiment, there were two separate slits. What was he trying to accomplish was demonstrate a wave function to light.

Today we have learned that light acts as particles, which is physical, and a wave that isn't physical. Much of it is dependent on the person who is observing to look. An

116

easy way to grasp that when it goes along a particular path, it becomes physical.

If it travels through multiple paths, it is not physically. Its physicality also alters based on the observer's ability to observe. When there's no one to observe it behaves like waves in a non-physical component. While it's being watched the wave solidifies into reality and experience and is transformed into physical.

One way to think about it is when you're on more than one route, that is that you have more than one wave , and the waves interact with one another and there is interfering patterns. When you drop a small pebble into the pond, it will show ripples.

When you drop two stones simultaneously you'd notice two ripples. In the end, the ripples will meet and the result would be something like the sacred geometrical structure of the pattern of the flower of life. I'm gonna go into a little bit more detail here in just a second. Particularly, you're likely asking yourself what, in this moment what does this relate to Mandela

Effect? Mandela Effect? I'll be weaving all this up in a moment.

In order to connect these concepts we'll use the concept of cause and effect that is linked to quantum entanglement. Basically the principle is "The Butterfly Effect," that, in its simplest form it basically states that a butterfly's wings could result in an entire typhoon around the globe.

In every action, there's an equal and opposite reaction , therefore, you're faced with a range of choices and options but eventually you've exhausted all possibilities and choices.

To not go too far from the topic However, in the realm of time travel, also known or referred to in science and fiction as time continuums, you're experiencing an actual reality. This reality is what's known as an experiential real-world reality that is experienced through time.

It includes a beginning, a middle point, as well as an ending. The Bible for instances of this. There are prophecies in the Bible which, until the moment is accomplished, x, Z, and y won't be fulfilled.

The freedom of will is the most precious gift God ever gave, and with the free will of mankind, we can possess the power to alter the course of events. The power of free will is our ability to alter the direction of events predetermined by God's timing God's beyond time. In the actual fact, Jesus says, "I am the Alpha and Omega" this could mean that Jesus Christ himself is time.

It's a fascinating thing to consider and what I'm looking at is that there are many kinds of continuums that time may be a part of. This is a complicated concept for a human mind to grasp and yet it's not all that difficult.

If you are faced with a option among "a" and "b" when you select "a," well, what they'd say in a different reality or universe you actually picked "b." There are only two choices that you can choose from "a," option "b." What kind of scenario would you have out if you chose the option "b" in place of "a?"

Everyone has those events in their lives in the back of their minds, when they reflect on the decision-making process and ask, "What would've happened if I'd done this" There's no way to know and you cannot alter the past, however you can alter your life the present and, by making a change today, using The Butterfly Effect, you can positively alter the course of your life.

In bringing it back to Mandela Effect, I wanted to highlight a movie known as The Butterfly Effect. It's an exploration of time travel.

When he learns that the possibility of doing this, it's a more unique than typical sci-fi genre where the protagonist can achieve it using his mind. However, when he alters something in time, this alters the timeline of his life. When the character returns to his present moment, the timeline shifts and the whole film is essentially his attempt to return to the time when he was in the past that is where he came from.

Every single thing he does seems to make things get more difficult. It is easy to see and an understanding of how one choice could alter the course of different events in a manner that you cannot see or even feel.

The reason I am mentioning this is that there was a show called "Falling Skies" which was concerned with an alien invading. . What was interesting about the show was the overlord aliens coming to invading. "A" They possess an hive mind "b" they're invading host bodies, and basically assimilating any lifeform to their hive mind changing the genetics of the people.

One of the things the villain states, at one point almost imply that the mind of their alien race was working at such a high level that they could sketch out the Butterfly Effect of cause and impact on a global scale.

According to the imaginary realm that all species out thereis aware of the existence of the causes and effects that is the Butterfly Effect on a quantum level using their high mind. I found it fascinating, particularly in the implications are related to the subject we're going to discuss in the in the next.

Chapter 7: Cern

We'll now begin by entering CERN. CERN is the name of a research center located in Europe for nuclear Physics. It also studies particle physics that is connected to quantum physical.

They are basically conducting tests. The majority of people are aware of LHC, also known as the Large Hadron Collider or the LHC. It actually has smaller accelerators. What these accelerators do is capture particles and then smash these particles into one another. When they collide, they become one another.

The response is that it releases specific amounts of energy. what they do is examine the collisions as well as the reaction and the energy released from the reactions. In essence, you can claim that what they're studying is determining the cause and impact that results from this Butterfly Effect of what happens when the particles collide.

It is interesting to note that they use Shiva to be one of the sources of inspiration.

Shiva can be described as one of the Lord of the dance. He's not discussing River Dancin', either. It's the destroyer of universes, or the destroyer of all worlds. The dance is a destruction the world, waiting for Brahma to return to rejuvenate or create the universe or the entire world and so, very intriguing icons they've selected for themselves.

So far as I can tell from my research information, and which I consider to be a bit limited colliding a particle with another particle isn't able to open the way to a portal or star gate. It can't destroy a universe. It's not capable of creating an unimaginable black hole this is from what we know , and also on the claims they make about what they're doing.

In the, I'm not sure what number of terabytes they've released on the data they've recorded What's the information in it except for their scientific language or perform math, or without being able to interpret what the information could be telling us what they're doing and saying that they're opening a portal, a stargate, or connecting to other dimensions.

We're not certain, and until someone has a look at every piece of information there is no method to tell. I'd leave the possibility of different possibilities but I would be cautious about going too much into speculation.

I'll say that once you are in quantum connection and how everything is linked to each other, it is a good sign that they're doing things on a very small scale and we can tell that they are, it will depend on the results they're discovering and the experiments they're playing on at a scale that's not accessible to the civilian population.

You're required to be at a government level or at an academic level to have access to these labs of research or, specifically, to access the information. The amount of money invested in this research that individuals will be able to protect their intellectual IP and intellectual property. Do with it what you like.

The scientists who are interested in subjects like that of the Big Bang Theory and the God particle is familiar with the nature of CERN is, what it does and potential. Without a conventional term to define what it is in essence, it's an Atom smasher. It's a lot more advanced than that, however, and gives scientists the possibility of re-creating what was the Big Bang that created the universe. It is a current and controversial endeavor in the name of science.

CERN is an acronym in French acronym which translates to Conseil Europeen for la Research Nucleaire, or in English CERN stands for CERN is the European Laboratory for Particle Physics located in Geneva, Switzerland. It took $9 billion to construct and is 17 miles in length, and is buried 575 feet beneath on the Earth to be able to protect it from radiation contamination from the surface.

What has created, and continues to make, CERN a focus of global interest is its attempts to develop theories that have the potential to change the life of the world. CERN can generate the equivalent of a gravitational pull 100,000 times the force of Earth and speed up protons to less than the speeds of light on its path.

If this is a bit scary, think about the factors that make scientists like Steven Hawking and religious leaders equally concerned about CERN's future initiatives. Hawking said in his autobiography Starmus in which he wrote the book "The God particle found by CERN could end everything in the universe." Other plans were believed to open the doors to another dimension for just a tiny fraction of a second but not enough time to allow an idea of the other dimension. But this was disproved from CERN in its web site.

One of the facts that you should be aware of regarding the workings that is part of CERN is the fact that it can take months boost the acceleration speed of CERN's particle accelerator in order to perform a single test. It's not likely that there are multiple experiments within a few weeks. Additionally technologically CERN is the whole facility in which these experiments are happening and it is the LRC also known as the Large Hadron Collider is the technology that is actually doing the job.

Technology with this huge potential is not without its critics as well as those who believe that it is Biblically associated with the Antichrist who is bent on destroying the world. There's a conspiracy theory that states there is an obvious symbolism of the occult in the facility's logo and in the facilities, which is believed to be practiced by a few members. CERN's answer to these allegations are that conspiracy theories will continue to be a hot topic for those that come up with the latest trend to get their hands on.

However, the most recent research to were conducted including the most recent of which was conducted in April of 2016, have produced no doomsday theories predicted by religious or scientific experts. The LRC is scheduled to shut down in 2019 in order to modernization (which is odd considering the current capabilities). The destruction-of-the-universe theories and opening-the-gates-of-Hell scenarios have not occurred, leaving CERN as a fading memory for most of the people of the world. Despite all the fear and angst of highly respected experts, CERN is expected to close with a bang.

The technology that is so dangerous and controversial as the LRC ought to be getting more attention in the media other than the tabloids. The idea that is circulating on internet is that Internet has it that Mandela Effect, where facts and the past are "misremembered" could disqualify the LRC from study of advances in science for several years up to the time that CERN is able to get operational. The general public could be affected with this Mandela Effect, but it is believed that a portion members of scientific communities may be affected, too.

In the end, it's something of a dilemma. On one hand, you can see some of the biggest discoveries in science over the past few decades being attributed to CERN. However, on the other hand, you have prominent physicists like Steven Hawking maintaining that its continuous use could be a catalyst to devastate the universe. Then there are the religious leaders who oppose the LRC's efforts because it could provide a way into the realm of Hell.

It is safe to assume that the next LRC will be more efficient after the upgrade, considering the advancements in technology that have occurred over the past decade. However, what is needed is scientists who will continue be curious about the possibilities for the LRC and who are committed to studying the science of Physics. The latest results are mostly uninteresting and haven't piqued the curiosity of young scientists. Future prospects for CERN as well as the IRC is not just dependent on the enthusiasm of the scientific community, but also from countries willing to provide the financing to ensure that CERN financially sound.

Chapter 8: Matrix/ Simulation Theory

There are some things in life that are simply way too complicated for people to grasp, and then there are some issues that are coming out as facts and people don't understand the full extent of it.

We'll talk about this matrix, which is in all of us. Are you aware of this amazing film The Matrix. It was among the films that opened people's eyes to the reality that their lives could be more complicated than what they think. The movie could actually be the result of a simulation. That's right. We're going to talk about simulation theory.

It's not just a YouTuber discussing it. It's not just one man who's fascinated by these topics. I'm going to discuss the results of one PhD one after the other PhD following an additional Medal of Science winner. The people who have spent their entire lives trying to discover the significance of their lives discovered that

130

there is something in the universe's structure Computer code, astronomy and much more.

Did you see that scene from The Matrix? I'm not sure if anyone has seen it. There's a character named Neo and he believes his life is normal, then suddenly it turns out the world he's in is sort of alternate reality. And there's a great scene where Morpheus basically says to the character.

Matrix quote

Neo: Right now, we're in an online computer program?

Morpheus: Is it really that hard to believe?

Neo:This isn't real?

Morpheus: What's real? What is real? When you're discussing things you can feel, it is that you smell or what you can taste or see the real thing is just the interpretation of electrical signals in your brain.

The scene you're watching is something you must be aware of and should provide you with a sense of hope since toward the conclusion of this chapter you'll find out what this means for all of us. Reality is

only the result of a simulation. This is in line with the notion that things such as the paranormal, such as the Mandela Effect, things like even Nibiru and the Anunnaki,
I believe Einstein did it the best way when he stated, "Reality is merely an illusion"
In the 70's, I'm not sure if anyone have ever seen the film, Blade Runner. The film is based on a fantastic book. Sci-fi author named Phillip K. Dick. In the 70's Philip K. Dick was on the cover and declared.

Philip : While many believe they can remember their the past, but I believe to have a memory of the experience of a completely different, and a very different current life. I haven't met anyone who has claimed this before However, I do believe that my experience isn't singular. The thing that is perhaps special is the fact that I'm willing to share my experience with others.

We live in a computer-generated reality, and the only way to know it is when some variable is altered and an change in our world takes place. I'll even say that that this impression could be an indication that at a moment in the past the variable was changed however, the program didn't function and as a result an alternate reality branched off.

Then, they laughed at the man. Now, there are many other big brains, creating big-time news, such as Elon Musk, who ... said... suppose cannot live up to what he said. had said that.

Elon: I remember the most compelling argument that we were in a simulation having been in a simulation, which I believe to be the following. Forty years ago, there was Pong with two rectangles and dots. It was the game of the day. Forty years later, we're playing 3D-realistic photorealistic simulations that have millions of gamers playing at the same time, and the quality is getting better each year. Soon, we'll have virtual or enhanced reality. If you believe in any kind of advancement that is the case, then games will appear to be identical to reality.Given this, we're on the right track to create games that can't be distinguished from reality. These games can be played on the setup box, or on an PC or something else and there would likely be billions of machines or setup boxes. This would suggest that the probability of being in reality is one in billion.

What they say is that if it's impossible to discern the difference, it's as real. It's just as real. The thing Elon Musk said was that, "Look, bottom line is that if technology continues expand, it's known as Moore's Law and that, technology will advance within the next 30 years we've arrived at the point we are at now. what's to come in the next 30 years?

There's more than one. There was a piece of paper that was composed and published by a person named Nick Bostrom.

The paper he wrote was written by him and the title was Are You living in the world of a Computer Simulation? In the paper, he laid out the three possibilities.

First, there is a chance that civilization isn't likely in the same way as it is today. It will never attain the status of technology. It's going to be destroyed before it does.

Another possible scenario is we're in a simulation , but we've simply stopped making simulations and we're now creating simulations from simulations

Third probability, which he states, "It's impossible to say that this isn't possible."

There are some of us who may exist in a different dimension and the rest of us are simply simulations that have become conscious and do not even realize they're just simulations. We'll dig into the Mandela Effects that everybody is so fascinated by. Why do some people experience them , and why are certain people not.

But he's not the only one. We also have Bostrom. We also have Elon Musk. We also have Philip K. Dick. We have lots of people, but we now have Google AI's Ray Kurzweil.

Ray Kurzweil: First it's not obvious that there's any difference between a simulation and a real. Sometimes , I speak about how the possibility of simulated human intelligence. Then some philosophers will come in and claim, "Well, that's a simulation and not real." In the event that you can simulate something that is as exact as the real thing I don't believe ... There is a distinct difference.

A different possibility could be that the universe is actually running on computers because scientists have looked at the laws of nature as being a series of computerized processes. It's quite plausible to think that our universe is an electronic computer. We've been taught at least in Western philosophies, that something is true about the universe is that fundamentally it's nothing more than a bunch of things that's particles and energy. It's real physical things. It could be that the stuff represents information but in essence it's just something else.

Perhaps, in the end, it's not really something, it's actually information. The things we refer to as electrons, particles and protons are simply representations of information structures. the fundamental element of the universe is information.

We've all heard of the phrase that the universe is constantly expanding. As with a computer program it is possible to continue to program and make it. It's being built at a rapid pace, and the funny thing is that while it's being developed, it's also being watched.

The more we look at it, the more we can see how far it could move. It's like we're creating. This is akin directly with people of faith who have stated since the beginning that we are all children of God and that we're co-creators both here and now, based on this model of simulation.

These people who were adamant about the concept of a creator are suggesting that we might co-create and are living in the creation by a higher being, or at the very least one that is more advanced or, if the opposite is true, that we are in the future and we're simply reliving in which we played the game.

Our consciousness is almost connected and is now living in of this system, much like Neo was. The most interesting thing is that there are limits. There are boundaries within games played on video, and there are limits within a computer program.

At present, they're trying to disprove it , because this concept of the simulation is so convincing that they're trying disprove it. In their efforts to prove it, they realized that it cannot be proven since there are limits.

Let me try this alternative method. When you gaze at a TV screen, you will see pixels. Right now, but the further you go, the more it becomes discolored. Then you get down to the "plank" level, which is the lowest point. What happens? There's no meaning. It's not logical in any way. It doesn't even exist. There's nothing in fact which is why there's more evidence to support the theory of simulation.

Quantum Physics is even saying that we live in a holographic world where is actually only linear code. The information we're streaming through is projected onto an Hologram.

I'm not saying we're on an extraterrestrial program. I'm not saying this even. I'm merely saying that experts are saying and there's definitely some reason behind it, and we need to look into the issue and give us some sort of enthusiasm because, if that's the scenario, then if you've ever seen the film, The Matrix, when Neo discovered the truth, the sky is the limit.

We are at the edge of gaining an almost god-like power over our minds and matter. There's an even more important moment in consciousness than at the present. It's only 20-30 minutes away from having the ability accomplish this. It's 300 million people at times the duration of the time in the Universe.

If this is an example of a simulation then all you need to write is a small piece of code, and then you'll see the tiny modifications. What do we then become? We're like lab rats. A kind of alien experiment. I'm not saying it's true, but it's definitely something we could think about because things are changing.

Chapter 9: Have We All Died ?

In the event that you glance at someone and tell them that nobody has ever really died , and in addition, nobody ever will, the person is likely to be looking at you like you were three heads. It's a reasonable response, as anyone you mention this to will likely think you're insane. What is the truth but? There are people who don't really believe that death is real. Before you decide to reject this concept, it's crucial to conduct some research regarding the truthfulness of the assertion first.

Christians believe that, while physical things happen but dying of soul is not something that occurs. There are other religions that share similar convictions. If this is true then you must think about how the belief that death is not a reality is so easily dismissed. Many refer to this notion in the form of death theories.

It is the truth that it's basically saying lots of the same idea that a large number of spiritual people already understand. It's putting scientific thinking to the concept that humans are only energy that manifests in a specific form. It is possible to argue that our souls are composed of energy.

It is the essence of your vitality and defines the person you are. The physical body does not have any connection with it. It is possible to consider it in the same way as you might imagine a box that houses an electrical device. It is the energy that is available when you think of the electrical device but the box is just an empty storage container. Also, your body.

Even Einstein acknowledged that the human body is composed of energy. It has been proven scientifically throughout a number of years that energy is not destroyed. So, it is reasonable to conclude that when the body of a person is gone, the energy that is the basis of a specific person actually moves to a new kind of existence on a different level.

Some have attempted to discredit this claim by referring to"the Mandela effect. In essence, they claim that those who claim to have experienced previous life events are reliving events that have never occurred to them. This is also true for people who claim to have ex-body experiences at the point of death. They assert that this is due to that Mandela effect,

People are reminiscing about an event that has never happened to them, mostly because this idea has been created in their minds from someone else. typically those who wish to be able to recall this event. This is why doctors, psychiatrists, and relatives have claimed to have caused this Mandela phenomenon that occurs. It's not enough to account for every single instance someone describes in the event of a previous life regression or an experience that is out of body. Thus, it's negligent to automatically believe that each memory someone experiences is necessarily wrong.

What can a person decide when they are in two distinct different groups that think differently? The ones who believe death doesn't happen are certain that they are right in their belief. In the same way, those who want to blame everything on a false memory is similarly convinced that their beliefs are accurate.

This is a problem to anyone seeking more details because every party tries to show what they believe is the correct method to tackle the issue. In reality, there's much about life that isn't known and by its nature, the only way to gain understanding is death.

Absolutely, every living thing does undergo a physical death. This can be proved by simply taking a look at someone in this state. Physically, they don't respond to any stimulus and are not adept at moving their bodies, making words or expressing thoughts.

But, it doesn't mean that there's no existence. The spirit of the soul is still present. It could be an issue of whether or not the person who is so desperate to determine what the finality of death is really aware that when physical death takes place the life continues in a totally different world. For some are able to accept, while for others, it's almost impossible.

In other words in the event that we die physically our consciousness or mind goes into a different universe that is similar to this one, but with a slight difference. If we die due to old age, then we are brought back in the form of a baby. If we pass away asleep nearing the middle of our lives, maybe we don't even realize we've passed away and get up in new bodies.

Doesn't it What about calling it Reincarnation. Imagine that you're returning to other realm. In the future, you might be happily married, with plenty of money and children but in the next you're a down and out laying about with 3 or four marriages that have broken down.

Our job may be to be able to experience everything that there was or could be. That means from having one kid to many children. To be shot dead by every bullet ever fired during the world war 2. To live in the past and to become President of the USA (I'm British, so I'm not eligible to be president at this time). In one lifetime, you'll be living a simple life in a small city, and in another you're supposed to travel around all over the world.

There are so many possibilities that this theory can bring, I'll let you experiment with it and come back to me.

Chapter 10: Land Mass Movements

One aspect of human life that has been the most affected by the Mandela impact is the changes in land mass. I'm not sure why the changes in land mass have occurred and why, but here's an overview of the reasons .

New Zealand. This is among the regions that was hardest hit by Mandela effect. Many people can vividly remember locating it on north of the North East coast of Australia. They are shocked when they realize that it's located in the southeast and south of Australia. Some of the people who are affected The Mandela impact in New Zealand, also seem to believe that the Australian landmass has grown significantly and appears bigger than they imagined. There are a lot of internet users who are able to find New Zealand exactly where it is, yet place Australia further towards South New Zealand. Many even recall New Zealand as being west of Australia.

Trinidad and Tobago. It is an Caribbean island with around 1.5 million inhabitants whose Capital city happens to be Port of Spain. There is a section of people who believe that the island is farther to the east. They claim that the island is moving closer to Africa. Some argue that it has now been essentially touching Venezuela. People who place Trinidad and Tobago close to Venezuela appear to place Cuba further away from the Yucatan peninsula. Their experiences seem strikingly like. A large portion of people who put the island there think it is further away from Granada and farther east in Barbados than it actually is. Its location in the peninsula doesn't make much difference either.

South America Continent. Many people seem to be convinced it is the case that the continent South America has drifted drastically towards the east. They claim that it's named "South" America for a reason. This is because it's directly underneath North America. It's now Southeast America, according to the maps. The reason they claim this is that Rio was previously within the same zone of time as Oklahoma. In reality, Oklahoma along with the City of Rio are not part of an identical time zone. This is why they believe that the continent is moving further east towards Africa further away from North American continent. Divers in Howard AFB in Panama claim that they were once traveling north-south on the canal. Today, they travel east-west.

Japan. For many people who watch the world, Japan was once further than South Korea with cities like Fukuoka and Hiroshima being a long way away from South Korean coastal cities like Busan and Daegu. It is believed that the Jeju Island was always considered more to the west and slightly further north than Japan. Presently, Japan seems to be quite closer with South Korea and further north than was thought previously. There's a dispute over the name that is brewing between Japanese as well as the Koreans regarding which part of they should call the Sea of Japan should be the "Sea of Japan" or the "East Sea." In the photographs where

Japan is shown as a single entity the amount of water we see on the map gives the impression that the island of Japan are surrounded by a large amount of water around it. When viewed in relation to other countries, the distance shrinks significantly. This could be an reason the reason Japan has been moving to the to the north.

Alaska. The Mandela effect may also have a negative impact on those who are able to take taken a quick look at Alaska. The most striking geographical changes are the two huge bays currently along the western coast of Alaska. Many people will claim that they are not in the past universe/reality. Another impact on Alaska is its size. It is a fact that the state of Alaska is currently five times its size in comparison with the 48 other states! Alaska's west coast Alaska hasn't always been as it does today. Another significant change in geography to Alaska is the result of Seward Peninsula. Seward Peninsula which many claim was never always like that.

Bermuda Triangle. The Bermuda Triangle, located southwest of Miami is full of mystery and riddles. The majority of these questions and puzzles focus on bizarre creatures and disappearing vessels and planes which cannot be found. But the Mandela influence on Bermuda triangle is based on the many contradictory information about the area. For a long time, it has been interpreted as starting out of FL, Bermuda, San Juan, Miami, and Puerto Rico. A lot of new photos currently show the region being engulfed by those islands in Puerto Rico, and Bermuda and the entire region in South Florida with the end point being in the Gulf of Mexico. The Mandela effect on the location of the region together with the eerie supernatural phenomena makes the area an ever-growing fascination for those with an fascination in that Mandela effect.

India. The Mandela effect isn't limited to people who live in India's West as well as those in the Far East. There are many who believe that India's subcontinent India was previously more triangular in its shape.

150

The huge subcontinent that is home to over 1 billion people is believed that it has lost land along the eastern shores. This is most likely due to the regions in Nellore south of Chennai and east of Madurai and Digha south of Kolkata. It is also believed to have been pushed farther north towards the Himalayan Mountains, particularly in the area of Himachal Pradesh. There's a substantial population of people who place India in the vicinity of as the Middle East such that Kerala is only a couple hundred miles away from the eastern coast of Oman.

Sri Lankan. First map that was drawn of Sri Lanka came into existence in the year 1535. It was drawn by Claudius Ptolemy. The mountains shown on the map have been significantly moved off towards the west when compared with current maps. The best way to grasp how the Sri Lankan Mandela effect is to look at the map from a an orientation of North-South. The following maps from Sri Lanka would surface nearly 170 years later, in 1650. The map illustrated Sri Lanka (Ceylon) with its surrounding waters as well as several tiny islands. There is a stark distinction in the position in Sri Lanka relative to India from the maps prior to the year 1700, and on maps that appeared in the latter half of 20th century. In all earlier maps, Sri Lanka (Ceylon) was depicted on the southeast of India near Chennai and further in India's Bay of Bengal.

Madagascar. Malagasy (or Madagascar as it's now known was first introduced to world time in the time that Marco Polo visited it by accident in during the thirteenth century. It's absurd that the country didn't have an official name in the indigenous Malagasy languages, despite being in existence from 350BC until 550 AD. This means that it isn't just the issue of maps by itself and also has an issue with habitation and an odd background. A lot of people have admitted that they see Madagascar as being closer to the African continent and farther in the south than it is currently. It is also believed to be a widespread and widely accepted belief that the island was more round. Some have stated that it has drastically changed with respect towards its proximity to Galapagos Islands.

Arctic. A lot of maps from recent times show it is the Arctic is merely a huge large mass of water. They show the Arctic ocean as a massive large blue mass, rather than depicting that the white glaciers the majority of which are melting. Many people believe that there was a land mass on the northern hemisphere, where they believe the Arctic Ocean is. In the majority of time, it was believed that the Arctic Ocean was frozen. It was represented as a white layer. The Mandela effect is due in part to the way that people believe that underneath the ice, there is a land that can be used for living and is able to hold it up. It's sometimes difficult to imagine large areas of solid, icy ice floating across vast expanses of ocean.What can make the Mandela effect even more pervasive is the fact that in the south of the Antarctic regions that had caps of ice were generally on top of the terrain. So, once the melting of the ice, maps simply reflected the erosion of land.

It is the Baja Peninsula. The Baja peninsula lies located in northern Mexico and is located between it with the Gulf of California and the Pacific Ocean. A lot of people, even those who attended school at the University of California and the regions surrounding the Sonora Region in Mexico don't remember drawing it on the maps they used in class. Many say that the rise that forms it to the Baja Peninsula either wasn't there or was too small to be apparent. Some say it is true that Baja California used to be totally thin, with no protrusions of that kind. To further reinforce this Mandela effect there are claims that of the "Sea of Cortez" wasn't there at all. It was completely part of the Gulf of California. There is also a group which claims it is the Baja peninsula was composed of two peninsulas, as opposed to one.

The Koreas. The Korean region is comprised out of North Korea and South Korea. They are bordered by the Yellow

Sea to the west, China to the northwest, Russia to the Northeast and the Sea of Japan to the North as well as the east and southeast. Many people are aware of The Korea being North of Japan and farther north than where it is now located. The entire area seemed to be an area smaller than it is today, with Okinawa, Philippines and Japan together between the north and south of Indonesia and South of Korea (note that the Philippines is farther south). Many people believe that Korea wasn't always in the south of Russia in the way it is now. Many people do not even remember North Korea sharing a border with Russia. Many who are able to locate their respective Koreas in a chart assert that they've always been slightly towards the right.

Greenland is more than twice as big. Greenland always appeared larger than it appears today. Actually, many maps showed it to be about twice the dimensions of the continental United States. Many people remain of the opinion that Greenland is among the biggest nations. One of the main reasons for this situation is poles are hard to accurately render from a sphere onto straight lines. It is the reason you're likely to see different renderings regarding the size and orientation of poles are related. Mercator renderings always exaggerate the dimensions of Greenland. It creates the impression that it is as large as Africa. Sometimes it even shows Greenland to be higher up than it is currently.

Cuba. The majority of world watchers believe that Cuba is moving further to the west, to its Gulf of Mexico. Many people are now of the opinion that Cuba is much larger and spread out more than they thought. Many people are seeing Cuba as a tiny , isolated island near on the shores of Florida. Cuba when seen from a distance is huge and heavily populated. In reality, Cuba is the biggest Island in the Caribbean and is nearly larger than the entire State of Florida. If viewed in relation with America, Cuba looks small and small. The majority of people are amazed to learn that Cuba can be described as an Orca whale-shaped islandthat is nearly twice as large as previously thought. It is a bridge that connects to the Yucatan peninsula to Eastern Caribbean. Cayman was always thought to be located further east and north from Cuba as its Caribbean Sea occupying a larger surface area that lies between Cuba Honduras and Nicaragua.

Sweden/Denmark. Because of the Mandela effect for a long time, a lot of people have not realized the fact that Norway and Sweden share a border and Sweden share a strait Denmark. Sweden and Norway appear to be in a arc, and this has greatly reduced their distance. However, the truth is Denmark as well as Sweden have such a close relationship they share a bridge connecting them , which takes only a few minutes to connect. There is a belief that Denmark is further south than Sweden and the northern part of Denmark pointed towards Stockholm. This is only some hundred miles from the present geographical layout. Skagen in the northern part of Denmark is much more close to Gothenburg located in Sweden. The majority of people have believed it was located near Grimstad which is located in Norway. They also think that there are more islands within the Oresund Strait which divides Denmark as well as Sweden than there actually are. In reality, many people mistake their Oresund Strait region with islands in the west of Scotland. Conclusion. There's no doubt that certain of the geographic Mandela phenomena are far more real and interesting than others. One thing that's not doubt in this instance is the fact that some of these effects are difficult to explain even after removing Mercator projections. In time, the majority of Mandela influences on geopolitics will eventually be clarified. The few that appear odd that need to inspire more research. In particular, the explanation of the brain for the Mandela effect is sure to improve as more research in this field are completed

Chapter 11: Bio Of Fiona Broome

Fiona Broome is widely known by many for her writing and as a researcher who specializes in issues related to ghosts, alternate histories and Faeries. The popular name for her is The Mandela Effect because of her passion for reading to the degree that she came up with an idea and named it "The Mandela Effect". One of the first stories she ever wrote was sold in the early the 1980s by Fate Magazine. Since then, Fiona has been true source of accurate information in different fields of extrasensory research.

Mrs. Broome has written a many great books that can easily be sold. She also has published numerous writings online and contributed to numerous magazines. She has ties with many publishers like schiffer books, Usborne, Sterling, West side and publications international.She has contributed to several anthologies that deal with ghost stories.

She has also been involved in a variety of Television and Radio shows as consultant, and she used her extensive experience to develop the celebrities of these companies, including SkFy, The history channel and Discovery. Fiona is also seen in a number of radio and television shows, including Hollywood New England and of course, the history channel. para-X Radio programs, Darkness Radio, Blog Talk Radio, Psychic Sundays, Doctor Rock and Witch, as well as the spiritual view.

Fiona Broome is also cited in the most viewed newspapers and magazines like Haunted Times in 2008,NH Magazine in 2002 and 2010 Celebrate magazine in 2010, the Boston Globe Newspaper in 2002 and also the cover story of Nashua Telegraph.She is also mentioned by numerous researchers in research documents, as well as other works of art.

In addition to her appearance at the international conference Fiona has also participated in investigations and at public events. She is actually one of the most prominent investigators in a few localities of The United Kingdom, the United States and Canada. In addition, she has research locations like Hounas House, Tenney Gatehouse, USS Salem, the Standing Inn, myrtles plantation, Hawthone hotel and several other sites for research.

She was also the creator of Hollow Hill, the oldest Internet site that was associated with ghosts, and earned her more respect online. The site was initially called HollowHill.com however, in the present it's changed to EncounterGhosts.com.

In the past few years, Fiona has earned more interest because she came up with and spread a theory referred to by The Mandela Effect which was an analysis of alternative reality.This theory suggests that a majority of people, particularly strangers, can recall numerous identical events that have similar specifics. However the memories we have aren't exactly the same as what is recorded in archives, newspapers and other history books. It's not about fake memories.

Fiona has also written more than 11,000 articles for magazines and websites and has also been a part of numerous international conferences and events as a panelist and a speaker. Some of these events are GhostStock the annual Ghost Conference held in Canada, New England ghost conference, and the Central Texas ghost conference in central Texas.

The year 2003 was the first time Fiona attended The New England Ghost conference where she was the keynote speaker for the event. She was invited for five years as the chief guest and speaker at Dragon con to educate the public about science fiction, art , and the paranormal. She was one of the panelists, speakers and investigators panelists at Ghostock 7. Between 2008 and 2009 she was the main attraction as well as host of the premier social eventin Salem in Massachusetts and Massachusetts. Fiona was also among the speakers and guests in the Central Texas paranormal conference which was held in Austin in Texas.

She is also highly regarded for her insightful suggestions and advice that have been gaining importance in the field in the last few years. In 1989, she released an article in Fate Magazine, Fiona's first piece under the name"Margaret Brighton". Her knowledge and experience in this field led her the perfect character model Fiona Character held in Trickery Treat.

Fiona is the top Paragenealogist of all time across the globe. She has a wealth of expertise in this area.Using the knowledge she's acquired over the last thirty years, she's capable of using sources of historical and genealogical research to help explain why certain places and people are more prone to supernatural activities than other places and people. The late Ms. Broome is very instrumental in explaining the paranormal and the historical facts that will explain the reasons for their existence. Since 1990 she has taught Ghost photography.

Chapter 12: A Listing Of The Top Popular Modifications

This is a brief, but not extensive of of the more known developments that are published. Keep in mind that I've listed the changes that have been reported from different sources which includes those that I could not have a connection directly to their authors. You might which is why we have listed these as well:

A lot of people believe Nelson Mandela died while incarcerated in prison, however Mandela was later released and later became Presidency of South Africa.

"The Monopoly Guy no longer has the Monocle. Newspaper articles continue to reference him as the Hasbro's "Monocled Mascot". We're not comparing him to Peanut. Peanut!

It is the North Pole no longer has an ice cap. It is the only place Greenland is covered with ice. However the two Prince Harry as well as Top Gear's Top Gear team were there and have pictures and video proof of ice in the area. And ask yourself what place does Santa reside now that there's no ice?

Six passengers in the car of JFK at the time of his assassination in lieu of only four. But an easy Google Images search for KENNEDY MUSEUM CAR can provide enough evidence to help you make your own choice.

The phrase "Luke you are my father" the words that Darth Vader spoke to reveal his identity to his son, no has any meaning anymore. It's now, "No. You are my father." There's a way to pull an old VHS version of the film from the garage, but you will not have "Luke". However, despite a lot of persons (even James Earl Jones himself!) and movies that joke about the phrase for years in the past as well as Google Images stacked to the edges with speech bubbles that are"Luke "Luke" version there are still those who are unable to accept that there is something wrong.

Interview with a Vampire has been republished in THE Vampire. You can grab that old copy off your shelf and your cover would have been changed to THE.

In the film Apollo 13, Tom Hanks mentions the famous, "Houston, we have an issue." Someone pointed out that the movie was changed to "Houston We had an issue." After they looked into the original NASA footage, they found that it was also changed to had, however, the film had been reverted to having. This could be an attempt to determine whether they could reverse the modifications they'd made. If you consider it in a rational way for a moment What would "We had an issue" mean? And why not abandon a mission if there's not an issue anymore?

- Volvo is a brand with an arrow pointed towards the upward direction to signify its male representation. A lot of people are aware of the it was never there However, if you look into the history of the logo you'll discover that it has always been this way.

-- Coca Cola is also a logo that has always been the way it is today even though most people will remember when it was long lower, lower and wavy like the symbol.

-- KitKat was Kit-Kat to a lot of people, but today there's no way of proving that the dash was ever existed in the logo.

In the well-known art work "American Gothic" The woman in the painting looks sideways, where the majority of people always was looking straight ahead.

" Terminator 2. Judgement Day is now, Judgment Day. According to Google it is an US or UK spelling distinction in the 1600s even though the majority of Americans spelling it as Judgement. If you were If a UK resident who was raised spelling Judgement throughout his life, he would have recognized the spelling distinction and have laughed about it as an imitation or misprint. The thing that makes this change interesting is the fact that the word is Judgment found in the Bibles of your children, and other films and books as well. We've asked many Americans about the meaning of the word and for every one who says that it's Judgment there are 10 spelling it Judgement.

In the Phil Collins song, "In the Air Tonight", the lyrics HOLD ON were replaced with the words OH LORD. (For excellent reminiscence, see the film The Hangover where Mike Tyson listens to the tune).

The film The Candidate starring Will Farrell and Zack Galifianakis changed to The Campaign and the Robert Redford movie is now named The Candidate.

The cereal Rice Crispies is now Rice Krispies.

The Ford logo has never featured the pig's tail, which is in the F.

As we've mentioned in the book earlier that this book has noted that the Volkswagen VW logo has now an arc that runs that runs between"W. This is a good reason to imagine that this was an uncomplicated logo change however, it nearly impossible to find even old images that do not have the line dividing them.

-- Fruit Loops apparently never existed. In this timeline, it's always Froot Loops.

The most loved beer, Bud Lite is now Bud Light.

The scene at the end of James Bond movie, Moonraker is a girl who smiles at Jaws, the villain. Jaws to show her braces. Since the two are able to relate to one another and are able to connect, they both run off together. However, if you were to watch the film today there's no movie that you can find that shows that the girl was wearing braces. This would negate the whole purpose of why Jaws could even discover an ancestry in her in the first place.

- The Capricorn was once the name of a Goat however, it now is now a Sea-Goat.

Do you remember Forest Gump saying, "My Mama always said that life is like a chocolate box You don't know what you're going to receive. .'"? Now, his mom always claimed that life was as a box of chocolates. It's absurd! If life was like an assortment of chocolates What is it like now? Also, in the film when she talks to him about that she is dying she says "IS". The Making-Of video shows another camera filming Tom Hanks as he was speaking the line. in this video Tom Hanks clearly states, ".....life is as a box of chocolates ..."

As I mentioned, Sketchers is now Skechers.

The all-golden C3PO robot C3PO from Star Wars now has one Silver leg.

The band that was Johnny Quest is now Jonny Quest.

The Berenstein Bears is now The Berenstein Bears is now The Berenstain Bears.

-- Febreeze has been changed to Febreze.

-- Jiffy Peanut Butter currently the only Jif Peanut Butter.

The lazy Boy chairs have always been La-Z-Boys.

-- Mirror, Mirror on the wall in the classic Disney film is currently Magic Mirror.

A. Mona Lisa now has a smile, but it is believed that she didn't.

Chic-Fil-A was always Chick-Fil A, as per the company.

The A-Team van was entirely black with a red dividing stripe surrounding it. It's now a duo-tone colour with the bottom half of the van is black and the top portion is gray.

According to certain sources, the position of Earth's satellite within the galaxy has changed.

The movie ends with the movie, ALIVE two men venture out in search of assistance. They cross a river, and a man riding on an horse sees them. The horseman is no longer in the film.

Many people are of the opinion that "tank person" protester actually was run into and then was killed in the course of the tank.

The Laurel and Hardy's line "This is yet another fine mess you've put US in!" is now "This is yet another fine mess that you've put ME in!"

In the movie JAWS, "WE'RE going to need a bigger boat ." right now, "YOU'RE going to need more space in your boat."

The people who owned the Home Depot store are baffled by the fact that it is now THE Home Depot.

in the track, California Dreaming, the group Mamas & The Papas sing, "I got down on my knees and pretend to pray" which many believe to be "BEGAN to pray." This is evidenced by the recurrence in a variety of versions of the song, in which they make use of "began" in place of "pretended" even by major International groups like R.E.M.

Adult diapers for Depends have currently dependent.

Queen's "We are the Champions" does not end at the end of "...of all the earth." However, you only have to search the lyrics to find out why it is supposed to. There's a live version of the song that ends this way and we're convinced that they didn't modify it intentionally. They're using it to disprove the assertion of it being ME by claiming this is the same version was played on radio for a long time and that's the reason everybody remembers it ending this in this way. However, we have both owned The Greatest Hits albums, and we both recall the song concluding "of all the earth" in the discs. The song doesn't exist any more...

-"BRAGG'S Apple Cider Vinegar is now only BRAGG Apple Cider Vinegar. They made a mistake on one time on the label on the back. Check if you can identify it before they fix the issue.

Chapter 13: Unknown

I've had numerous encounters that involved parallel universes. At the very least, evidence that parallel universes are real.

At the age of seven in the first grade in the Catholic elementary school the nuns announced that they would be conducting auditions for a variety show featuring talent. I was an avid dancer and singer so I volunteered to be the role of the talent coordinator on the stage at school. I clearly remember singing the song from
My Fair Lady
. The lady wrote my name on her notebook and said "Very excellent. You'll be a part of the show's variety." The nun was thrilled! Then, I realized that weeks, days and even months passed and no mention about any show of variety. The variety show never took place. I'm not even sure I ever asked the reason why

there wasn't a variety show, and there was no anyone ever mentioned it. This was a bit unusual for me, as I'd be eager to be a part of this show and would have questioned what was going on. It's a memory I have to this day.

In the 90s, as my husband was browsing through photo files on his computer there were a few photos that I had taken as a kid that he had never previously seen, at least, not like those. I looked at them and then I saw myself at the time I was in second grade, wearing my uniform at school, carrying my book bag in the school's park but I couldn't remember the specific picture. I was pointing at something that was on the ground.

I'd never seen the picture previously, and don't think I remember ever taking a photo with my back to the ground

. There were some other photographs of me from on the same day that I recognized in the same uniform as well in the exact same place. I'd seen them hundreds of times before. But not the one in which I was pointing to floor! Then I looked over different pictures of myself and this time, when I was in the third grade living in my parents living room. Another picture that I was unable to recognize, as there was furniture in that living space that was completely different from the furniture we'd ever owned!

I was compelled to show these photos to my mom, who had always been the primary owner of all our family photographs. She was aware of every picture of me that had been taken. She could pinpoint exactly when the photo was taken, who was the person who took the picture, and from where. She'd never ever seen such pictures before, and justly, it was freaking her out!

After a few minutes the images disappeared from the computer but the pictures were not gone before both my mom and I were given many opportunities to examine them and attempt to determine how it could have happened.

The only way I am able to explain the reason for these images is to go back to what was happening at that my life. I had attended the audition and was accepted to the variety show, however, an anomaly took place in the timing pattern. There was a gap between the universes. This could explain why I had forgotten about the show (at at least for a short time) and never inquired about it. It was this universe that I was slipping into. This was the universe in which I never had my photo taken, and was pointing toward the ground. It was the universe where there was none of the variety shows, there was no audition and no reason to inquire about the show since it was never a reality. Later on, a few memories came to mind concerning my auditions for a variety show and I was wondering why I never inquired about the show. It was like there was a dark hole between the two memories, almost like it happened to two totally different individuals. In the world of

these strange images that had never previously been seen I
DID
Participate in the auditions in the talent competition and then
WAS
It was indeed a talent show. It was my finger towards the ground, I'll probably never understand however I do know there was a significant distinction to be noticed, as even though I looked exactly the same in the photo of me as a third-grader, in the living area of my home, I did not identify the furniture. A major shift in the reality was happening, as seen in two images. Two different realities, same person.

He's forgotten about the incident but I'm convinced that the portal he entered was into another world. The other cats may have noticed this, which is the reason they all gathered in that particular area beneath the table. There was probably one small gap which he tried to get through, which could be the cause of the scratches across his face. Beatnik likely poked his nose inside, in order to escape. The other universe like, we don't know. However, whatever that was definitely terrified him.

Portals similar to these that connect to other universes appear, occasionally. We don't know the way or what causes it, but they do occur. He got through, but luckily, was in a position to get back. He was definitely content to return.

In another instance, back in August 2012 my husband and me were in a state of meditation. Sometimes, we do this to increase the energies to more positive levels to generate positive results. We were engaged in meditation when something occurred. It was loud to the two of us. We didn't really know what to think of it. In that moment, however I could sense that something was different. First of all, I was planning to visit Trader Joe's in the afternoon and buy a bottle of wine. I had a habit of having an alcoholic glass with my meals, and I had run lacking wine. It was then that I realized that I was not going to purchase any wine. In actual fact, I was certain the fact at that point in time that I wouldn't ever drink wine or any alcohol beverage again. The world had been transformed in a radical way. In the afternoon we went to visit my mother-in law's grave, and we discovered that many things had changed about the grave. A large hedge that had been present was gone and there was no evidence of it ever being there. The headstone's name adjacent to hers was totally different from the one which was on the headstone before. I could tell exactly what it was before, prior to it being the same name as my mother's: Georgina. The name that was last is similar to one of my all-time favorite comedians from the past: Gleason. In the cemetery, the front desk claimed that the fence we spoke about was never there and there was no "new" title on the grave was always there. The next morning, our vehicle drove by the town where he was born, Glendale, California, and we noticed another oddity There was a church still standing, where it never been as well as a place that we had always frequented was gone. What had changed within the two weeks that had passed since the last time we visited? Weird.

Conclusion

My first prediction Based on data from the past like dates of operation in this LHC is another increase in the development of Mandela effects in the future phases of similar research. The amount of new Mandela effects will rise in direct proportion to the growth in the amount of data collected during these periods of time. For instance, the volume of data gathered during the future runs of the LCH along with its successor(s) will probably grow exponentially, which means the amount of visible Mandela changes will increase.

The process of adjusting to changes is likely to continue as rationalization continues to be a factor , regardless of the number of quantum changes that occur in a particular time period.

It will be harder to attribute changes to memory loss, if an increase of significant magnitude does actually occur during these periods of time as I expect. There could also be a slight increase in the appearance of Mandela effects in the run time because of an increasing amount of resources required for the maintenance mode of such a large amount of information. The increases should will be replaced by an increase of resources, however, there will be a few fluctuations in the emerging pattern.

Another hypothesis is that Mandela changes will also be more noticeable due to the volume of data that is gathered through these types of tests expands. The amplification in breaking Mandela changes could be fixed by hand, but those that aren't corrected will become more notable when the volume of resources diverted increases.

If my forecast is right that is the case, then the changes be more noticeable because experiments are able to gather more information . It becomes more difficult to write these changes as memory issues. If the tolerance to allow the persistence of the changes is at the point where subtractions are possible like Shazaam or Shazaam, then additional significant changes could be observed during the next surges.

It is possible that noticeability will be reduced by the frequency distribution of every type of quantum shift, since the substitutions will remain greater than subtractions with shifts with a frequency that is between the two. This buffer against noticeability occurs due to the fact that Mandela substitutions are much more easy to conceptualize than subtractions. A proportional share of every kind of Mandela alters relative to the total amount of Mandela effects that are evident in each spike is likely to remain constant.

www.ingramcontent.com/pod-product-compliance
Lightning Source LLC
Chambersburg PA
CBHW060330030426
42336CB00011B/1275